PHYSICS IN ACTION

Electricity 2
Electromagnetism

GORDON RAITT
Director, School Physics in Engineering Project

CAMBRIDGE UNIVERSITY PRESS

Cambridge
New York New Rochelle
Melbourne Sydney

Published by the Press Syndicate of the University of Cambridge
The Pitt Building, Trumpington Street, Cambridge CB2 1RP
32 East 57th Street, New York, NY 10022, USA
10 Stamford Road, Oakleigh, Melbourne 3166, Australia

© Cambridge University Press 1987

First published 1987
Reprinted 1988

Printed in Great Britain by Scotprint Ltd, Musselburgh, Scotland

British Library cataloguing in publication data

Raitt, Gordon
Electricity 2 : electromagnetism.——
(Physics in action)
1. Electromagnetism
I. Title II. Series
537   QC760

ISBN 0 521 31086 5

GK

The cover shows a technician completing a transformer (see page 43).

CONTENTS

Preface   iv

Narrative section

Engineers and electromagnetism   1
Powering and controlling some machinery   2
Some magnetic fields   5
Non-destructive testing: magnetic particle inspection   8
Solenoids   15
Controlling machinery: industrial relays   19
Electromagnetic clutches and brakes   22
Making alternating current electricity   32
Transformers   39

Development section

A   Basic electromagnetism   47
B   Magnetic fields and magnetic particle inspection   48
C   A working solenoid   49
D   Alternating current generators   51
E   Transformers   54

Answers to questions   57
Index   58

PREFACE

About this book

The main purposes of the School Physics in Engineering Project and its books are:
* to show ways in which physics that students study in their GCSE courses is used in practice
* to provide real numerical examples and problems for class work and homework
* to interest students in engineering.

These are central features of the aims of GCSE courses in physics.

The book describes a range of electromagnetic machines and an industrial test procedure – magnetic particle inspection – which between them cover the main work in electromagnetic fields and electromagnetic induction in GCSE syllabuses.

The book is intended as a complement to the physics textbook that is in class use in a school. It aims to show how physics that a class is being taught is used in practice. It does not set out to teach the physics although it provides revision summaries of key parts of syllabuses. The Narrative Section links electromagnetic machines and devices with physics principles and provides some simple problems. The Development Section extends the work and provides additional problems, including some numerical ones.

The book can be used in different ways. The Narrative Section can be used to introduce the class teaching of a syllabus section of physics; or it can be used after the teaching for emphasis of key points. The revision and problems in the Development Section can be used for class work or for homework. They can complement the questions in the class textbook.

The School Physics in Engineering Project

The main purposes of the Project were given in the previous section: "About this book."

The initial work was done in the Physics Department of a large comprehensive school from 1979 to 1983, with $7/10$ teaching and $3/10$ curriculum development programmes. Regular visits were made to construction sites, and visits were made to manufacturing companies. Modules were written, and were used with classes in three comprehensive schools. Evaluation returns from the schools have formed a basis for revising and extending the texts, and for developing them into book form. This last stage, which has included further industrial visits, was carried out during an attachment to the Department of Education of the University of Southampton, 1983–1985.

In order to ensure technical accuracy, each section of the texts together with its photographs and diagrams has been sent to several assessors for checking. These include a combination of site engineers, engineers in the companies whose machinery or plant has been described, polytechnic staff and university staff.

The work has been funded by the Industry Education Unit of the Departments of Trade and Industry, the Comino Foundation, The Cement and Concrete Association, Dow Mac Concrete Ltd, Paterson Candy International Ltd, The Precast Concrete Industry Training Association, and by a loan from West Sussex County Council. This funding made the Development possible, by providing release time during industrial working hours for the necessary visits, and by providing a final period of fulltime working to complete the Project. I am very grateful indeed to the sponsors of the work.

Acknowledgments

The companies which gave information about their products and which gave guidance in describing them are acknowledged under the appropriate Figures and Tables. I am very grateful to these companies and to their staffs.

Dr J. W. Warren of the Department of Physics, Brunel University, kindly read the draft text and his detailed comments have enabled me to clarify and improve a number of aspects. I am very grateful to him.

Table of units

Unit	Symbol
millimetre	mm
centimetre	cm
metre	m
kilogram	kg
tonne	t
second	s
minute	min
revolution	rev
hertz	Hz
amp	A
volt	V
kilovolt	kV
watt	W
kilowatt	kW
megawatt	MW
ohm	Ω
joule	J
newton	N

Narrative section

Engineers and electromagnetism

Electromagnetism is used for transmitting power, for controlling machines, for transmitting information, and for storing information. It is of great everyday importance and use in the home, in offices, in factories, on farms, in cars, in trains, in aircraft, in ships and in communications.

Electromagnetism is used in generating electricity, with an alternator, or dynamo. When the electricity has been produced it can be made to do work, by passing a current through an electric motor; and electric motors function by electromagnetism.

Electromagnets can be used to control machinery. They can switch circuits on and off, as the relay in Figure 1 does. The magnetic field of a coil, or solenoid, carrying a current can be made to pull a steel plunger; and this can pull part of a machine. Figure 2 shows an industrial solenoid and its plunger. The two eye-holes on the left are for attaching machinery to the plunger.

Electromagnetism is used in telephones; a telephone earpiece is worked by an electromagnet. Radio, television, and microwave communication links all use electromagnetic waves. These communications systems allow us to send and receive information, and they are all based on electromagnetism.

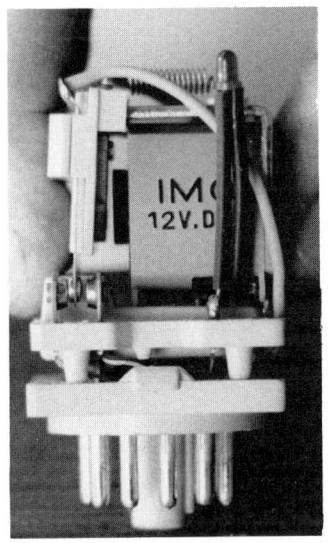

Figure 1 This is an industrial relay. It is an electromagnetic switch. The electromagnet is in the centre portion, under the cover marked IM. It is used for controlling machinery. The photograph is approximately the actual size.

Figure 2 This is an industrial solenoid. The central part is a coil of wire on a hollow support. The lefthand part is a steel plunger. It is used for making machinery move. The photograph is approximately the actual size. (Warner Electric.)

In most computer systems the storing of information in the 'memory' is done electromagnetically, and getting the information back again is done electromagnetically.

Before we can use these helpful machines and devices, they have to be invented, tried out, redesigned, remade, and tried out again. Then they have to be improved further, and made as production models.

Whatever sort of inventing, designing, testing, and making an engineer may be doing, he or she will almost certainly need to use electromagnetism.

Powering and controlling some machinery

Power and transmitting power

The machinery in Figure 3 is in two main units. On the left, with two silos, is a concrete-making plant. Just to its right is a machine which looks like a crane, but which is actually a 'boom scraper'. Both of these are powered by electricity. The concrete-making plant has four electric motors in it, and the boom scraper has two. The electric motors are worked by electromagnetism.

The electric current which drives the motors is produced by electromagnetism, in the alternator at the electricity generator.

When one of the motors is rotating, power has to be transmitted from the motor to where it is needed. In the boom scraper this is done by an electromagnetic clutch. Each motor in the boom scraper has its own electromagnetic clutch: to engage the motor with the transmission, and then later to disengage the motor and the transmission.

Inside the concrete-making plant, much of the machinery is operated pneumatically: that is, by compressed air. The compressed air moves pistons, and these move other machinery. The compressed air has to be controlled; it has to be allowed into the cylinder so as to move the piston, and then it has to be shut

Figure 3 The machines in this photograph are powered by electromagnetism: by electric motors, with electromagnetic clutches.

off. This is done by valves, and the valves are operated electromagnetically. Each valve is operated by a solenoid and plunger, like the one in Figure 2.

In Figure 4, the engineer is looking up at the operator's cabin. The operator cannot control the plant by rushing around it pulling levers. He sits at a control panel, and he works the machinery by remote control, using electricity and electromagnetism.

Control the machinery yourself: follow the operator's handbook

Figure 5 shows the layout of the operator's control panel. Try using it to make a batch of concrete. The operator's handbook gives the instructions.
1. Switch the plant on, at the mains isolator switch, A1.
2. Start the mixing motor, and mixing paddles. Press button A2, the pan motor start.
3. Start the compressor. (What will it be compressing: sand?/gravel?/cement powder?/air? Why will it be doing this?) Press button A3.
4. Weight out cement powder. Press button B.
5. Weight out water. Press button C.
6. Weight out aggregate: that is, sand, fine gravel, medium gravel, coarse gravel. Operate switches D1–D4.
7. Empty the aggregate into the loading trolley. Operate switch E.

Figure 4 The engineer is looking up towards the operator's cabin, with the windows. The operator controls the machinery with devices which use electromagnetism: a transfomer, relays, circuitbreakers and an electromagnetic brake.

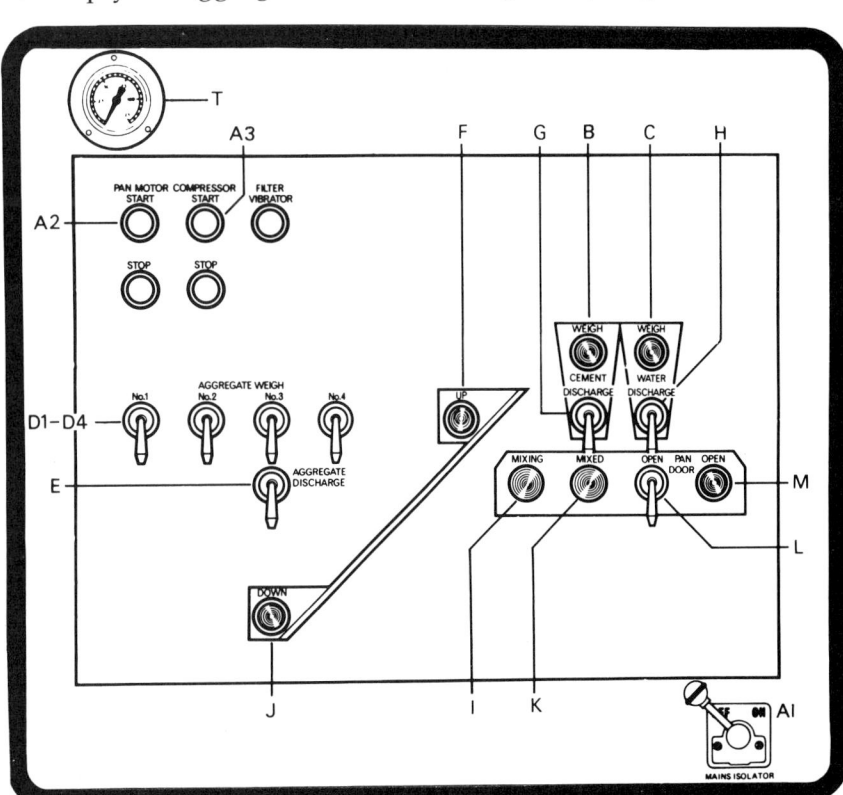

Figure 5 This is a diagram of the operator's control panel, on the concrete-making plant. (Benford Limited, UK).

3

8 Raise the trolley to the mixing pan. Press the UP button, button F.
9 When the trolley reaches the mixing pan it automatically tips the sand and gravel into the mixing pan. (Are the mixing paddles already moving round? Look back at operation (2).) Release the weighed cement powder into the mixing pan. Operate switch G.
10 Release the weighed water into the mixing pan. Operate switch H.
11 The timing device, T, starts automatically; and its **amber** light, I, comes on. Return the trolley to the sand and gravel loading point. Press the DOWN button, J.
12 When the **green** light, K, comes on the mixing is complete. (What do you think K is operated by? Look back at operation (11).) Release the batch of concrete into the storage compartment. Operate the pan emptying door, switch L.
13 While the pan is emptying, the light M shows **amber**. When the pan is empty, the light M shows **green**. Close the pan door; release switch L.
14 **Either** make another batch of concrete, and start again at operation number 4, **or**, close the plant down. You decide to have your lunch break and close the plant down. What operations will you do? (**hint**: Instead of doing operation (4), work backwards from there.)
(a) Stop the ----------. Press button --.
(b) Stop the ----------. Press button --.
(c) Switch --- at the ------------. Operate switch --.

Each of these buttons or switches closes or opens an electric circuit. When it does so, it is usually an electromagnetic device that operates.

Some of the electrical switch gear for the plant is shown in Figure 6. In the lefthand panel, on the top row there are five contactors. A contactor is a switch which is moved on and off by an electromagnet. It is used for switching on and off large currents, of over about 10 amps. The lefthand contactor is for switching on and off the air compressor motor, and it is designed to carry a current of 30A.

In the second row there are five circuitbreakers. They are electromagnetically operated, and they open a switch if the current in the circuit rises above a safe level. This breaks the circuit and protects the equipment from damage by too high a current. The lefthand circuitbreaker is for the air compressor motor. It is set to go off if the current rises above 15A. Later in the book you can read about how electromagnetic circuitbreakers work.

The third row contains nine relays, like the one in Figure 1. A relay is another type of electromagnetic switch. It is used for switching on and off currents of up to about 10A. Later in the book you can read about how a relay works.

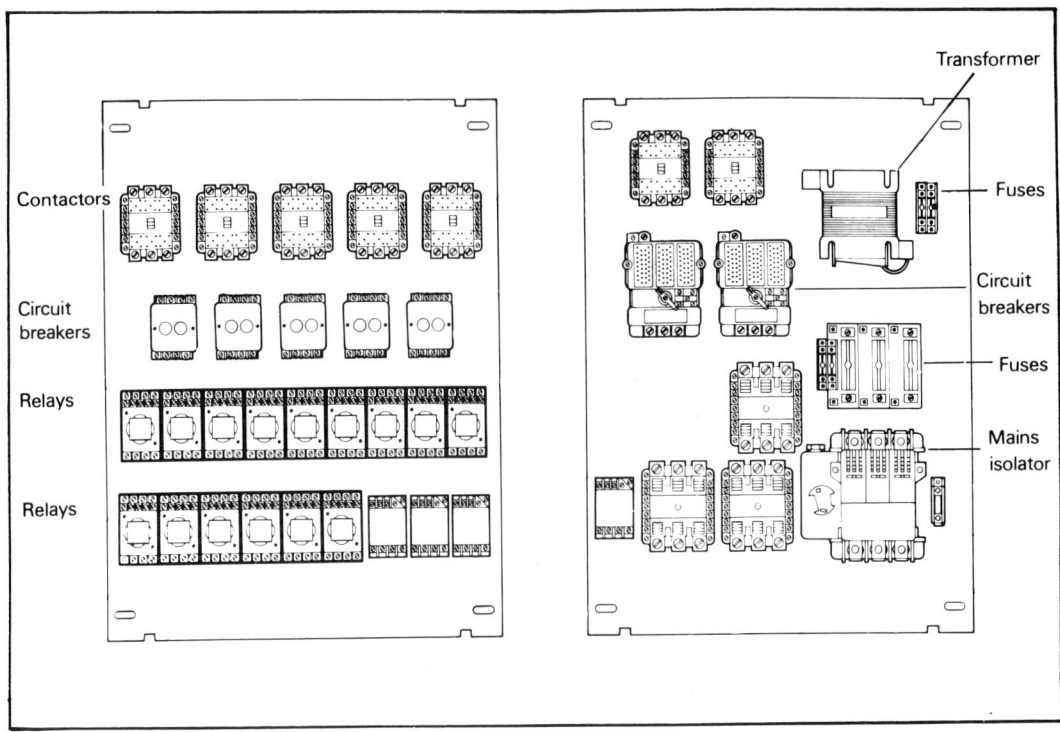

Figure 6 This diagram shows some of the electrical switchgear in the plant. Almost all of it is electromagnetic. (Benford Limited, UK).

In the righthand panel, in the top righthand corner is a transformer. This uses a 415 volt supply and transforms it down to 110 volts. The 110V supply is then used to work the electrical control gear, the relays and the contactors. Later in the book you can read about how transformers work.

Some magnetic fields

Figure 7 shows a straight non-magnetic conductor, such as a copper wire, carrying a current. The current produces a magnetic field around the conductor. If a small magnetic compass is placed near the wire, and if it is moved along in the direction in which the needle points, then the compass will trace out a circle around the wire. The magnetic field appears to be arranged in a circular manner about the wire.

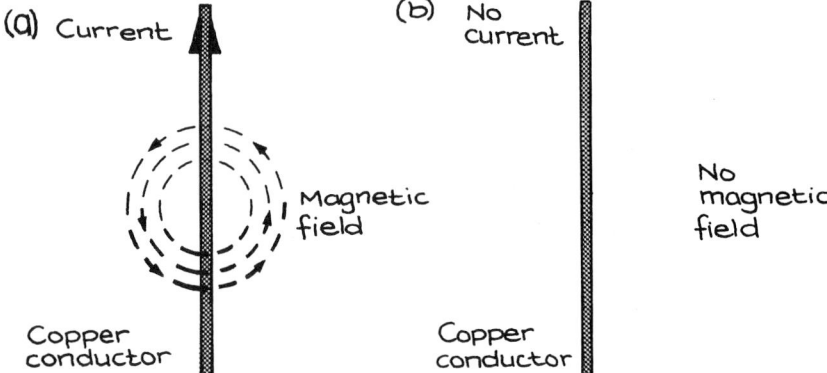

Figure 7
(a) This shows a current in a straight copper conductor, and the magnetic field which the current produces.
(b) When there is no current, there is no magnetic field.

If the current is switched off, the compass needle will not be able to detect any magnetism near the wire (except the earth's field). **The magnetic field was due to the current itself**. If the current is switched off, the magnetic field ceases to exist. If a current is passed along the wire again, a magnetic field again exists around the wire. Because the magnetism is produced by an electric current it is said to be electromagnetism.

If the conductor, say a copper wire, is coiled into a long coil and a current is switched on, the shape of the magnetic field becomes different. This is shown in Figure 8(a). The dashed lines which show the shape of the field are called **magnetic flux lines.** The arrows show the direction in which the north-seeking pole of a compass needle would point if it were placed in that position.

If the current is switched off, the magnetic field ceases to exist. This is shown in Figure 8(b).

Figure 8(c) shows the shape of a magnetic field around a permanent bar magnet. If this magnet is moved over to the coil in Figure 8 (a), the two magnetic fields will affect each other. If the N pole of the magnet is taken close to the N pole of the coil field, the magnet and the coil will **repel** each other.

If the S pole of the magnet is placed near the N pole of the coil field, the magnet and the coil will **attract** each other.

The magnet can exert a force on the coil, although **they are not touching each other.** Looked at the other way round, the coil can exert a force on the magnet, even though they don't touch each other.

Figure 8
(a) This shows a current in a coiled copper conductor, and the magnetic field which the current produces.
(b) If there is no current, there is no magnetic field.
(c) This shows the field around a bar permanent magnet.

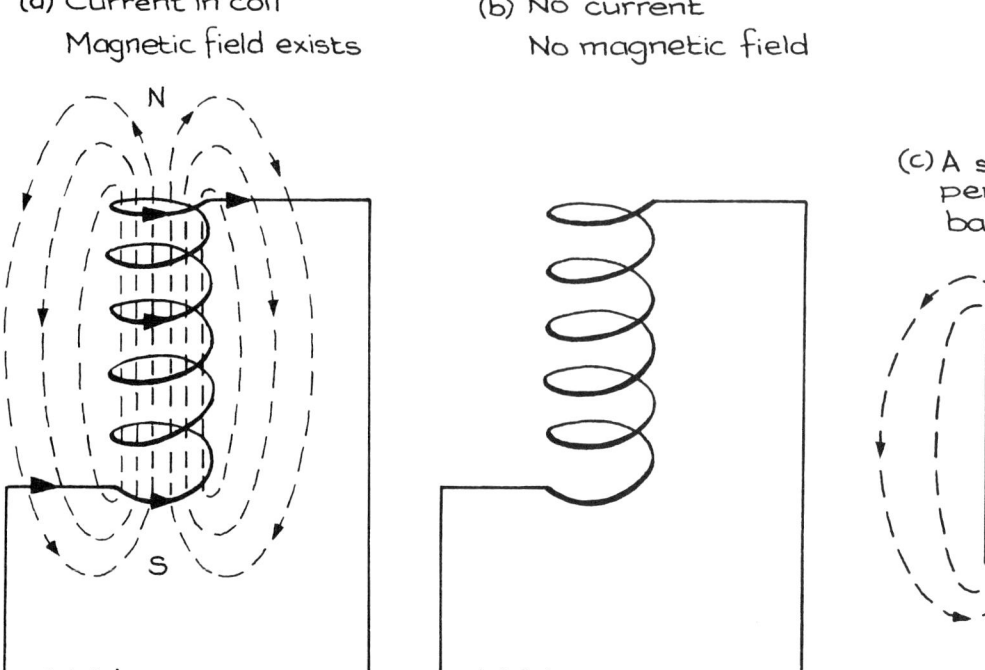
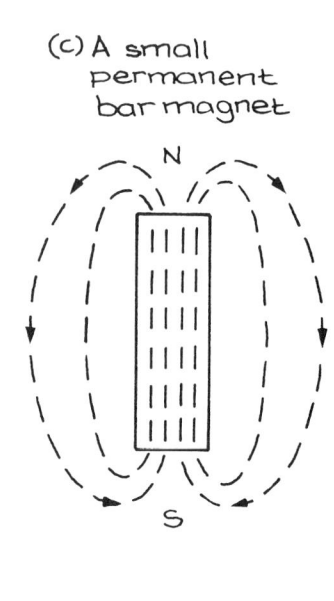

Magnetic fields can be used to transmit force through space. This is certainly something which engineers can use.

If a long bar magnet is bent round into the shape of a horseshoe, the shape of the magnetic field changes. Figure 9(a) shows the magnetic flux lines for a horseshoe permanent magnet. Is the top lefthand end of the horseshoe a N pole or a S pole?

A temporary magnet of the horseshoe type can be made by using soft iron, or magnetically soft steel. Figure 9(b) shows how this is done. Turns of insulated wire are wound round part of the horseshoe. When an electric current is passed through the wire, a magnetic field is created. The system becomes a horseshoe magnet. When the current is switched off, the field ceases to exist. This type of magnet is an **electromagnet.**

The importance of electromagnets is that they give magnetic fields which **can be switched on and off.** This makes them very valuable in designing and making a great range of machines and devices.

Figure 9 This shows some horseshoe magnets, and their magnetic fields.
(a) This is a permanent magnet.
(b) and (c) These are electromagnets.

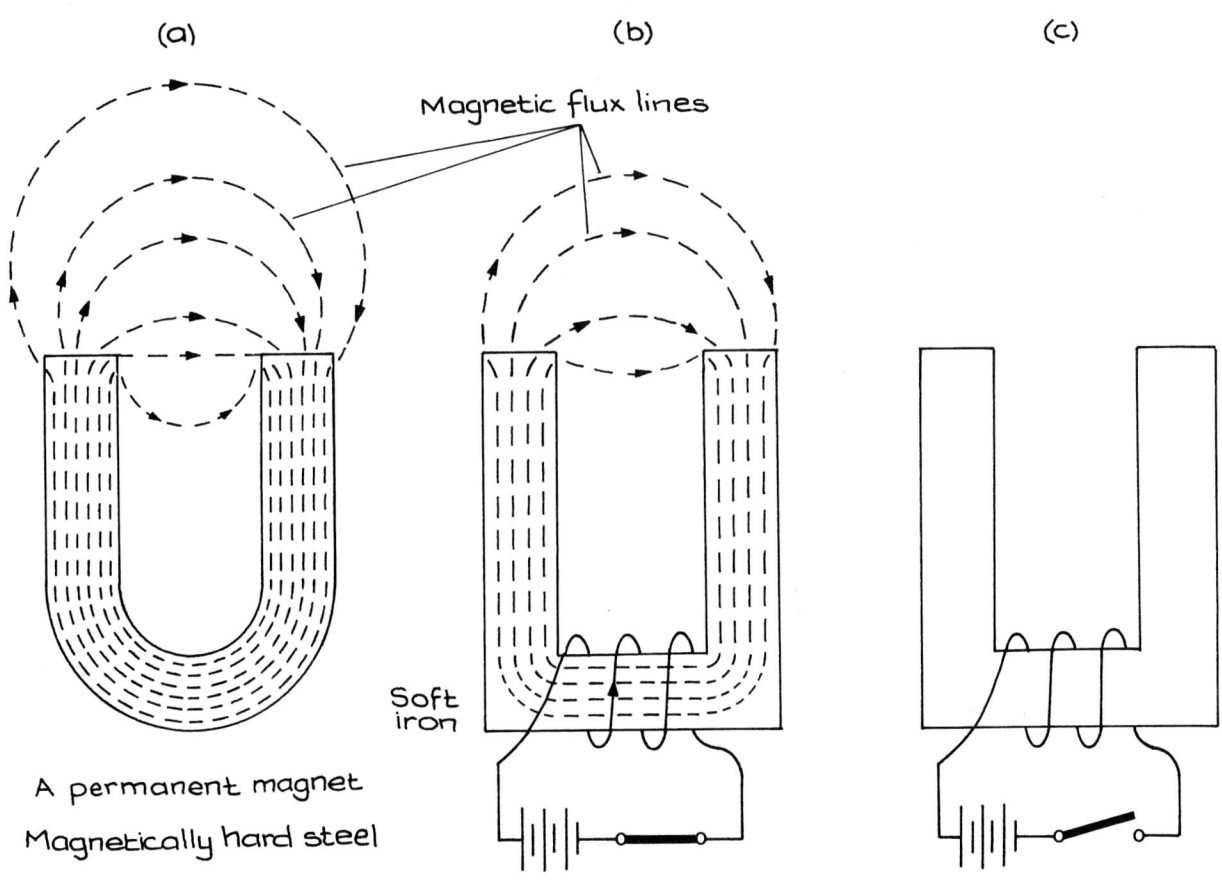

Non-destructive testing: magnetic particle inspection

What is non-destructive testing, NDT?

If you feel unwell, for instance with a bad cough, and go to the doctor, you would not expect the doctor to immediately get out some anaesthetics and scalpels and open up your chest to look at your lungs. You could expect the doctor to sound your chest with a stethoscope. If the cough has been there for a long time, the doctor might ask you to have a chest X-ray. He or she could then look in the X-ray photograph for signs of disease. The doctor is trying to find out what is wrong with you without doing any damage to you.

It is the same in industry. If a machine has become damaged, it is important to find out where the damage is and how bad it is without doing any more damage. If a component has just been manufactured, it is important to check that it is free from cavities and cracks that are not supposed to be there. It must be free from any fault that could cause it to fail in its job. Some ways of checking are needed that will not damage the component. This is called non-destructive testing, NDT.

The technician in Figure 10 is using an electromagnet and fine iron particles to look for cracks in the steel pipe. There are many different ways of doing non-destructive testing. X-rays can be used, to produce an X-ray photograph (x-ray radiography). Gamma rays can be used, to produce a gamma ray photograph (gamma radiography). Very high frequency mechanical vibrations can be used (ultrasonic testing). With ferromagnetic materials, like iron and steels, magnetic fields and magnetic particles can be used (magnetic particle inspection).

Figure 10 The technician is inspecting the end of a steel pipe for cracks. He is using an electromagnet, and he is spraying fine particles of iron onto the pipe. (Metal and Pipeline Endurance Limited, UK.)

When can defects develop in materials?

Washing machines, cars, trains, aircraft and other complicated machines are made by assembling together many different parts. The materials of these parts can develop defects, or faults, which could make a part unserviceable.

Defects or faults, can develop at different stages in the life of a part or component. They can develop right at the very start, for instance when a molten metal has been cast in a mould. When the metal has become solid and is cooling it shrinks, and cracks can develop. Figure 11 shows a billet of steel with cracks in it. The cracks have been shown up by fine iron particles coated with a fluorescent dye.

When a part is being machined it can be damaged by a bad machine tool. If the machine tool is not cutting cleanly it can tear at the part, and tear cracks can form in the part. When grinding is used to make a part smooth, poor grinding can produce grind cracks.

Even if a machine is in perfect order when it is delivered, defects can form later. They can form when the machine is in service. Increasing and decreasing strains on a part, repeated many times, can lead to fatigue cracks; then the part could break up, perhaps with disastrous results.

The fact that washing machines, cars, trains, aircraft and other complicated machines usually give good service for many years is a tribute to the care with which the parts are made and are tested at every important stage in their manufacture. Many machines are also regularly inspected and tested during service.

Magnetic particle inspection is the best and most sensitive way of detecting fine cracks and shallow cracks in ferromagnetic materials. It is very widely used throughout industry for this purpose.

Figure 11 This photograph shows a steel billet which has been rolled. Long cracks are shown up by fine iron particles coated with a fluorescent dye. The billet is about 125mm in diameter. (Magnaflux (UK) Limited.)

Magnetic fields and cracks

It is easy to form magnetic flux in iron and steels. It is more difficult to form magnetic flux in air. Magnetic flux will take a path in iron or steel rather than in air, if possible. Figure 12 shows this.

Figure 12(a) shows a permanent horseshoe magnet; and magnetic flux is spread out in the air space around the poles. In Figure 12 (b), a steel bar has been placed between the poles; all of the flux between the poles is now in the steel bar. None of it is in the air. It is easier for the magnetic flux to form in the steel bar than in the surrounding air.

If there is a crack in the bar, some of the magnetic flux escapes around the crack. There is a **flux leak.** This is shown in Figure 12(c). If very fine iron filings are sprinkled over the bar, then filings gather at the magnetic flux leak and are held there. If the top of the bar is blown gently, most of the iron filings are blown away; but those at the flux leak are held in place, and **they show up the place where the crack is**. This is how magnetic particle testing works.

If a machine part, or component, is ferromagnetic and if it is to be tested for cracks, this is how it is done. First, a magnetic field is formed in the part and around it. Then fine iron particles are sprinkled on, or are poured on suspended in a liquid. Then the iron particles are gently blown off, and the part is inspected by eye for iron filings held to the part. If there are filings left, they will be along any crack.

To show up the iron filings more clearly, they are usually dyed with coloured dyes. Often the dye is a fluorescent one, and if ultraviolet rays are used the particles glow brilliantly.

The positions and shapes of the cracks can then be seen clearly.

Figure 12 This shows magnetic flux lines in a good steel bar and in a cracked steel bar.
(a) A permanent magnet, producing magnetic flux.
(b) The flux path in a good steel bar.
(c) Some flux escaping near a crack in a steel bar, and iron particles gathering where the flux escapes.

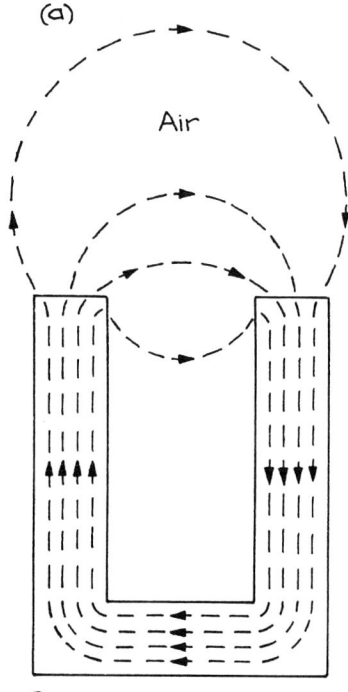
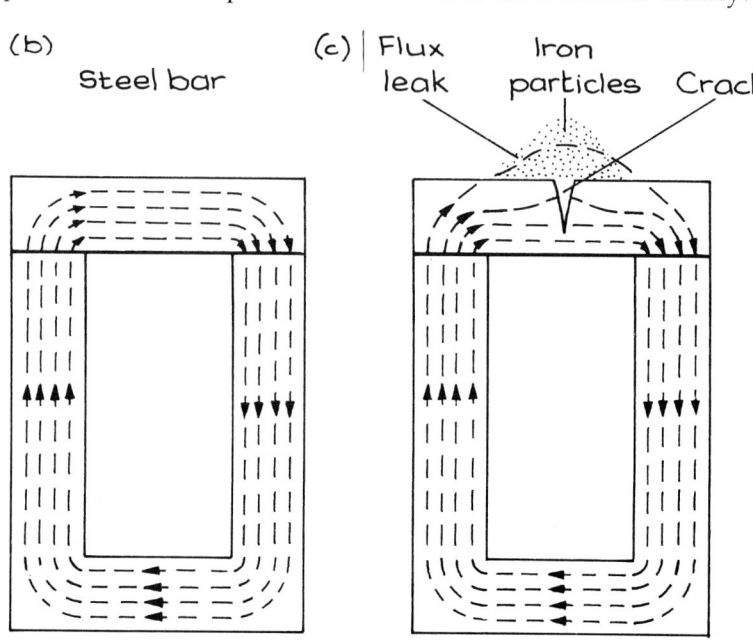

Magnetic field direction and crack direction

Figure 13(a) shows a steel bar which has been magnetized. The bar has cracks in it. Cracks which are at right angles to the direction of the magnetic field, like crack A, show up very well when treated with iron powder. Cracks which are parallel to the magnetic field, like crack B, do not show up. Cracks which are at about 45 degrees to the magnetic field, like crack C, show up slightly.

In Figure 13(b) the same bar has now been magnetized in a different direction. The magnetic flux lines show the direction of the field. If iron powder is sprinkled on the bar, will crack A now show up very well or not at all? Will crack B now show up very well or not at all? How will crack C show up?

What does this tell us about how a magnetic particle inspector should go about his work; how should he or she use the magnetizing field on the part being tested? Once, or twice? And in what directions?

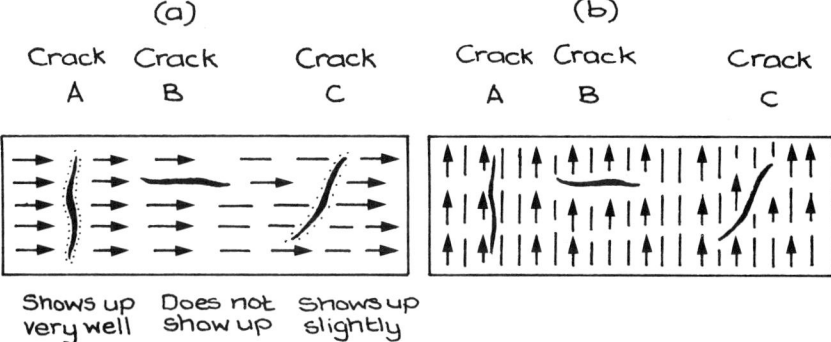

Figure 13 This shows a magnetized steel bar, with cracks in it. It is seen from above, in a plan view. The flux lines show the direction of the magnetic field. In (a) the field is along the length; in (b) the field is across the length.

Today's inspection

Job number 1

The photograph in Figure 14(a) shows an electromagnet that has been designed for detecting cracks in ferromagnetic materials. This particular model works on a 240 volt supply and uses a current of 1.5 amps. The top part is a handgrip. The model is designed to provide a strong magnetic field for crack detection. It is not designed to lift loads; but it would lift up a component of mass 10 kilograms. This gives some idea of how strong a magnetic field the electromagnet produces.

1. Why do you think that the legs are jointed? Have a look back at Figure 10, where an inspector is using one of these magnets.
2. What do you think the inside of the handgrip looks like? Draw a sketch of it, and label the parts.
3. Draw a sketch showing magnetic flux lines in the component when the magnet is on.
4. Figure 14(b) shows the cracks in the component being tested. There are seven of them, labelled A to G. With the

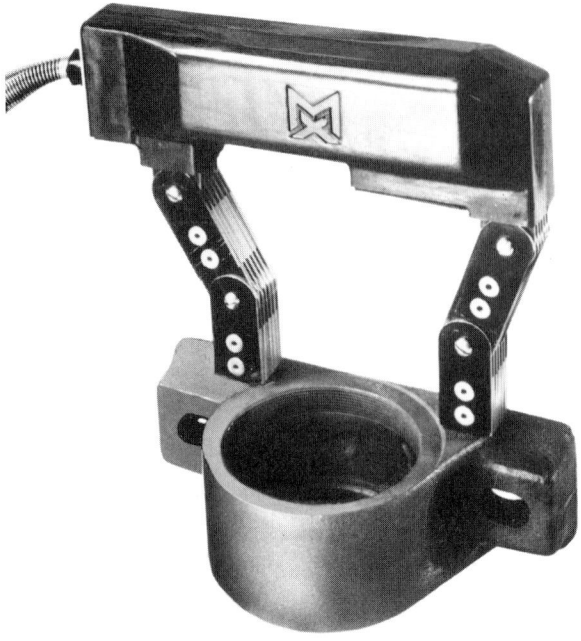

Figure 14
(a) The photograph shows an electromagnet which is designed for crack detection. (Magnaflux (UK) Limited.)

(b) The diagram shows a component with some cracks in it.

electromagnet in the position in the photograph, the component is magnetized. The magnet is removed; fine iron powder is sprinkled on the component; and the component is gently blown. Would the following cracks show up well: A, B, C, D, E? (You may need to look back at Figure 13.)

5 Sketch where you would place the poles of the electromagnet to detect a suspected crack F. Sketch in the flux lines.

6 Sketch where you would place the poles of the electromagnet to detect a suspected crack G. Sketch in the flux lines.

Job number 2

Your next inspection job is a steel cylinder, about 1 metre long and about 200 millimetres in diameter. This is far too large to be tested with the hand electromagnet that you used for Job number 1. Instead, the cylinder is placed between two supports and a large current is passed down it. Figure 15 shows the arrangement.

1 First, look back at Figure 7(a), page 5, which shows the shape of the magnetic field produced by a current in a straight conductor. The flux lines are arranged in a circular manner about the straight wire.

2 Now look at Figure 15 again. The cylinder being tested is a straight conductor. The magnetic flux lines are arranged around it in a circular manner. The cylinder has been treated with a fluid containing iron powder, and a crack has shown up. The line of the crack is at right angles to the direction of the magnetic flux lines, and so the crack shows up.

Figure 15 Another way of getting a magnetic field is to pass a current through the material being tested. The flux lines are in circles around the current path. (Adapted from Operator Guidance Chart, Magnaflux (UK) Limited.)

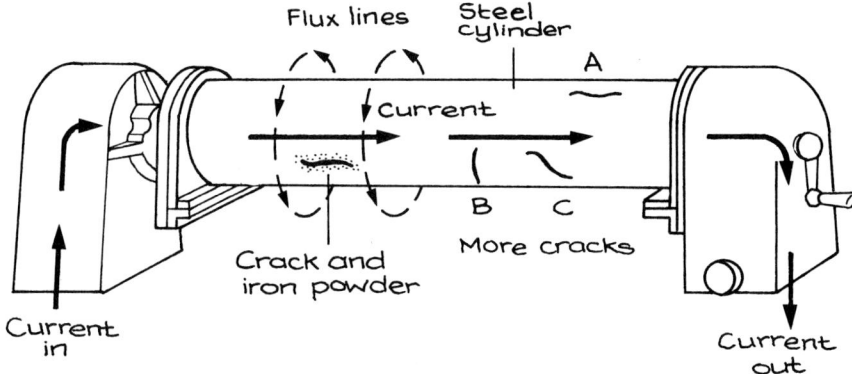

3 What size of current will you need: 3A, 30A, 300A, or 3000A? You could need about 3000A; this is a very common current for this sort of test.

4 After passing the correct current along the cylinder, you now inspect the righthand end. Which cracks will show up, out of the three cracks A, B and C? You could answer by using the words 'slightly', 'not at all', and 'very well'.

Job number 3

Figure 16 shows a technician doing a magnetic particle test on a component. It is an engine con rod, and it is supported on two end-pieces. On the right is a winding handle, to move the end-piece along. Just to the left of the handle is a very thick cable. It is bolted onto a copper bar which leads to the end-piece, and so to the component.

1 What is the cable for? Why is it very thick?

The technician is holding a hose and delivery tube. From the tube, a liquid containing dyed iron powder is flowing onto the component.

In the question about cracks A, B and C in the cylinder in Figure 15, you probably answered that crack B would not be detected. If so, you were right. The crack length is parallel to the magnetic flux lines. The situation cannot be left like this. If the

Figure 16 The technician placed a component between end-pieces. A large current was passed along the component. The technician is now using a stream of fluid with iron powder in it. (Magnaflux (UK) Limited.)

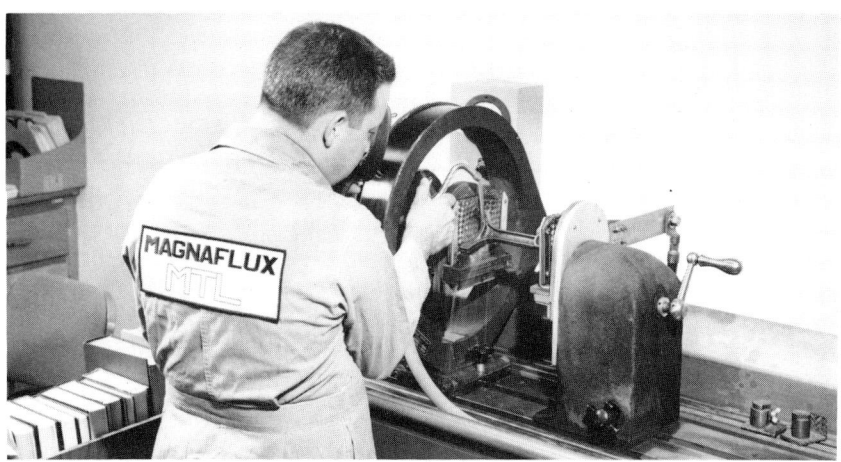

crack is there it must be detected; and there may be others like it. What is needed is a field with flux lines which are parallel to the length of the cylinder, that is, going from left to right, or right to left.

2. The way to get such a field is to place the cylinder in a coil of wire which carries a current. Figure 17 shows the arrangement. But, first, look back at Figure 8(a), page 6; this shows the shape of a magnetic field produced by a long coil of wire.

Now look again at Figure 17. The field produced by a short coil is similar. Close to the coil, the flux lines are parallel to the axis of the coil. This is just what is needed. The flux lines are at right angles to the crack; the iron powder is held, and the crack shows up.

So, cracks like A in Figure 15 can be found by passing a current along the specimen; and cracks like B in Figure 15 can be found by using a coil.

3. The equipment in Figure 16 can do both of these tests. The coil is at the lefthand end of the equipment. How is the coil mounted? Will the component be moved, or will the coil be moved, to get the component inside the coil?

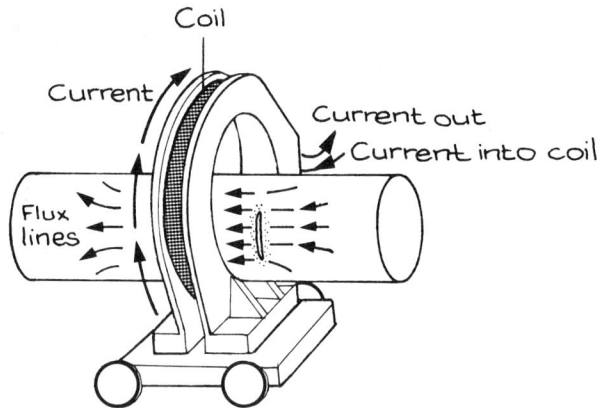

Figure 17 The diagram shows magnetic flux lines which are produced by a current in a coil. The arrangement can detect cracks that the equipment in Figure 15 could not detect. (Adapted from Operator Guidance Chart, Magnaflux (UK) Limited.)

Demagnetization

The magnetic field which is used to test components turns many of them into quite strong permanent magnets. Others are turned into weaker permanent magnets. For many purposes this is not desirable.

A magnetized part will keep some of the iron powder sticking to it. Other ferromagnetic particles, like small pieces of steel, may become picked up by the metal and will stick to it. These particles can prevent the part from being properly cleaned. They can affect painting; paint could later flake off over loose particles. If a magnetized part is machined, chips or flakes of metal can stick to the part and then interfere with the machining. The chips and flakes that stick can cause a poor finish to the part, and shorten the life of the machine tool.

Because of these reasons, it is very often necessary to demagnetize a component. There are several ways of doing this.

One method uses a yoke electromagnet, like the one in Figure 14(a). Alternating current, a.c., is used in the electromagnet; and the part is drawn through between the poles and then away to a distance.

Another method uses a coil, like the coil in Figure 17. Alternating current, a.c., is used in the coil; and the part is drawn right through the coil and away to a distance.

Solenoids

An electromagnet circuitbreaker

In physics books the word 'solenoid' means a long coil of a conductor, wound in the shape of a cylinder. One was shown in Figure 8(a) and (b). Have a look at this Figure and the shape of the magnetic field produced by a current in the solenoid.

In engineering the word solenoid means the coil and a cylinder of soft iron, or magnetically soft steel, which can move inside the coil. The cylinder can be attracted into the coil when a current is passed through the coil. The coil and cylinder are a way of producing motion. The motion can be used to move a switch, or to move parts of machinery.

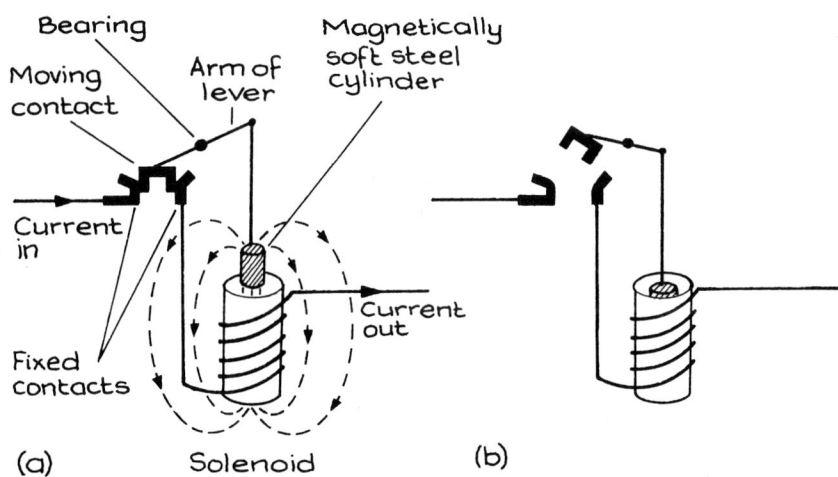

Figure 18 This is a diagram of a simplified electromagnetic circuitbreaker. It is worked by a solenoid. (a) Current flowing; (b) Circuit broken.

Figure 18 shows a simplified type of electromagnetic circuitbreaker. Circuitbreakers are very important in industrial circuits, and in domestic ones, for protecting machines and cables if the current rises to a dangerously high value. In Figure 18, the moving contact and the two fixed contacts are made of metal, such as brass. There is friction between the moving contact and the fixed contacts on either side, and for normal, safe currents the moving contact stays held as shown in Figure 18(a).

On a sheet of notepaper, describe how the circuitbreaker works **when the current rises greatly.** Describe what happens to the magnetic field, to the cylinder, to the lever, and to the moving contact. Why are there no magnetic flux lines in Figure 18(b)?

The arrangement in Figure 18 would work for a while; but after several operations the contact surfaces would wear. The friction force would become less and less. The breaker would go off at smaller and smaller currents, and it would become unreliable.

There are other problems, too. When a switch is opened in a circuit that is carrying a large current, an electric arc forms in the switch. Some way must be found for extinguishing the arc before it burns up the switch. After a breaker has gone off, the user needs to be able to reset the breaker simply and quickly. Designing and making a good electromagnetic circuitbreaker needs ingenuity and skill. Figure 19 shows designs being drawn for electrical equipment.

Most of the circuitbreakers used in industry are combined electromagnetic and thermal breakers. They contain a bimetal strip that is heated by the current. The strip bends as its temperature changes; and the bending can set the breaker off.

Figure 20 shows the inside of a breaker. The whole circuitbreaker can be held in a hand, comfortably. Find the current coil, the magnetically soft steel plunger, and the thermal strip. Also look for other parts. The arc chute puts out the electric arc.

If a circuitbreaker went off immediately the current went over the rated value, say 20 amps, that would be very inconvenient. In real circuits the currents change for many reasons, and they often go above the intended value for a short while. Circuitbreakers are designed to allow this. In the circuitbreaker in Figure 20, if the current becomes five times the intended value, the breaker goes off after 10 seconds. If the current becomes twenty times the intended value, the breaker goes off in one hundredth of a second, 0.01 s.

Figure 21 shows an assembler making a circuitbreaker unit.

Figure 19 The draftsmen and draftswomen are drawing designs for electrical equipment. (Crabtree Electrical Industries Limited, UK.)

Figure 20

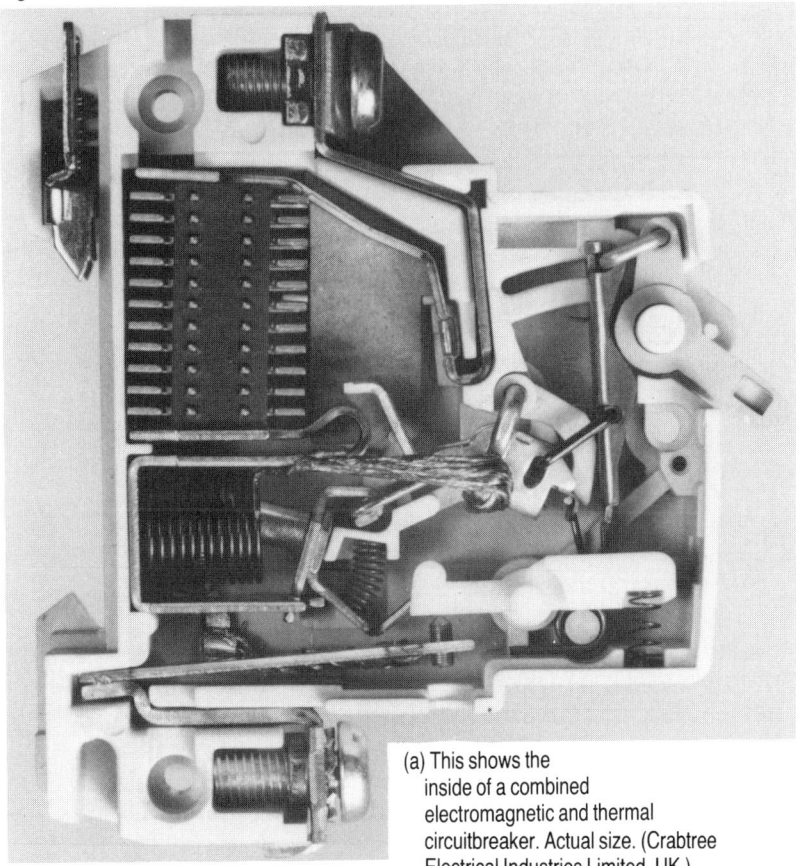

(a) This shows the inside of a combined electromagnetic and thermal circuitbreaker. Actual size. (Crabtree Electrical Industries Limited, UK.)

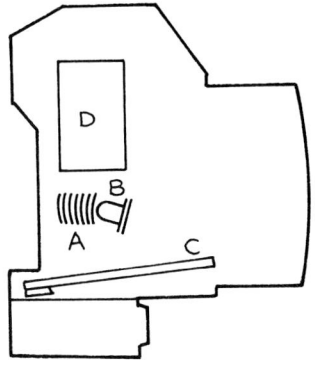

(b) Key to some parts. A, current coil; B, magnetically soft steel plunger; C, thermal bimetal strip; D, arc chute.

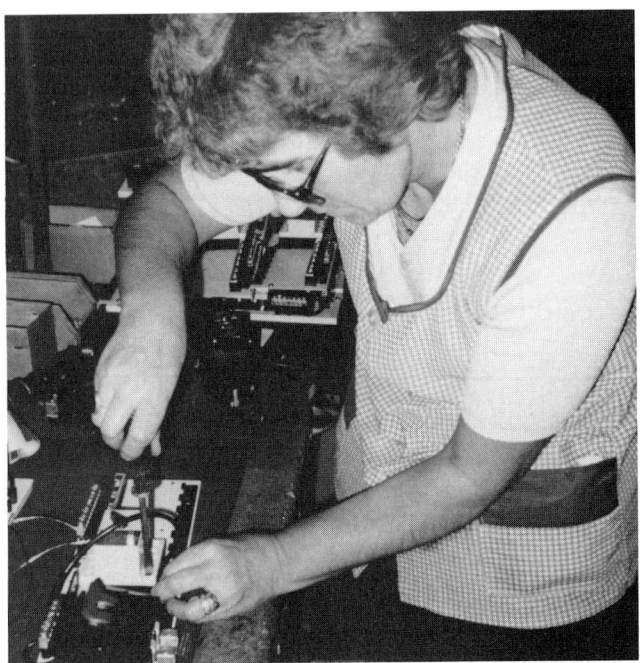

Figure 21 The assembler is making a circuitbreaker unit. (Federal Electric Limited, UK.)

An industrial solenoid: devise a working system

Have a look back at Figure 2, page 1. It shows an industrial solenoid. The plunger will move a distance of 25 millimetres and it can produce a force of about 40 newtons. Using a pulley and some string, what mass would the solenoid be able to lift? Give your answer in kilograms.

Figure 22 This is a sketch of an industrial solenoid. Compare it with the photograph in Figure 2. (Warner Electric.)

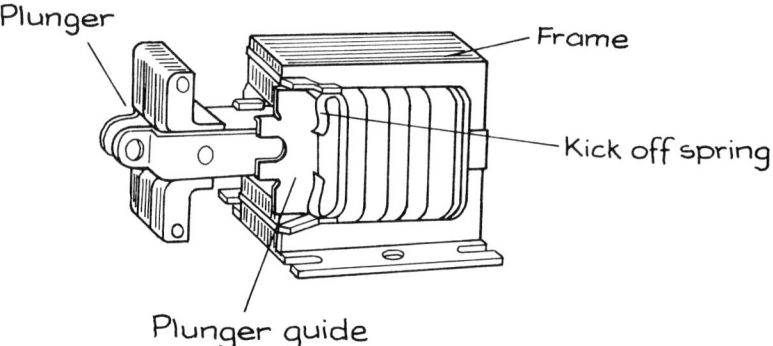

Figure 22 is a sketch of a solenoid. Where do you think the coil is?

You are designing a machine in which you use a solenoid to move a small lever. (1) Draw a sketch of how this could be done. (2) When the current is switched on, the plunger moves, and moves the lever. When the current is switched off the plunger remains where it is, inside the body of the solenoid. Devise a way of getting the lever and plunger back to their original positions. Draw a sketch of your system. (3) Devise a way of preventing the plunger being pulled out of the yoke. Sketch your system.

Another industrial solenoid: study its properties

Figure 23 shows that the force of magnetic attraction on the plunger of a solenoid varies with where the plunger is. The 'stroke' of a plunger is how far it can still move inwards. If a plunger is fully out, and it can move 20mm inwards, then its maximum stroke is 20mm. When the plunger is fully in, the stroke is 0.

1. For the solenoid, Model D3, what is the maximum stroke?
2. If the solenoid is at 20°C, and the plunger is fully out, and the power is switched on, what is the force of attraction on the plunger, in newtons?
3. The plunger moves in. When there is 8mm left to go, what is the force of attraction?
4. When the plunger is fully in, what is the force of attraction?
5. As the plunger moves into the coil, does the force of attraction become greater or less?
6. The manufacturer has provided two curves, one for the equipment at 20°C and one for it at 90°C. If the solenoid is being used in air that is at room temperature, what might cause the solenoid to reach a temperature of 90°C?

If the current is kept on for a long period, passing along the

Figure 23 This is a graph of how the force of attraction on a solenoid plunger depends on the position of the plunger. It is for two different temperatures. (Warner Electric.)

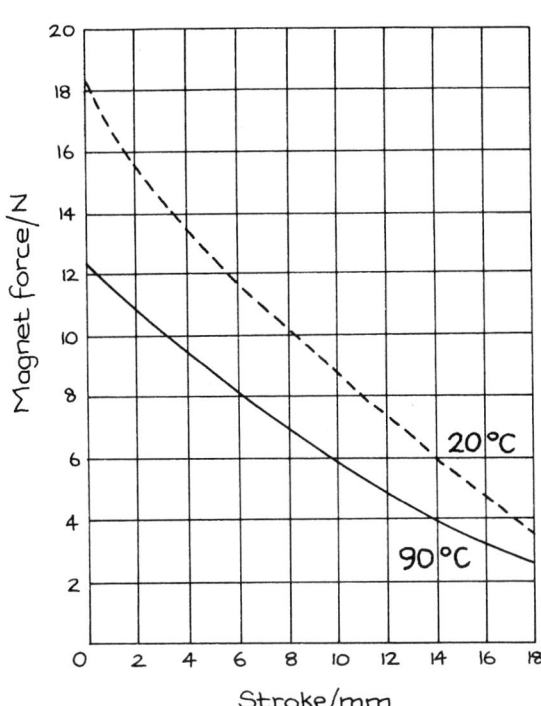

wire of the coil, it does provide a magnetic field for a long period. What else does it do? The current in the wire generates heat, and this raises the temperature of the solenoid.

7 When working at 90°C, is the solenoid more effective or less effective than at 20°C? For many uses, the designers of the mechanical system would have to take this into account.

8 Suggest and sketch some suitable way of allowing heat to escape from the solenoid: Bolt the steel base to a large steel plate? Use cooling fins? Use a fan? Some other way?

9 Write two or three sentences describing how heat is removed in your system (conduction, convection, radiation).

Controlling machinery: industrial relays

What is an industrial relay?

Have a look again at Figure 3 and Figure 4. The operator has to control electrical machinery which works at 415 volts and with currents of perhaps 20 amps or 30 amps. It is not a good idea to have the operator handling switches at 415 V and carrying 20 A or 30 A. One reason is safety; but there are other reasons as well, electrical reasons.

Now have a look at Figure 1 and Figure 24; they show a relay. The key part is an electromagnet.

A relay is an electromagnetic device. It allows an operator to use a **low voltage, low current** circuit to operate a **magnetic switch** in a **high voltage, high current** circuit. The high voltage, high current circuit has the electrical machinery in it.

The electromagnet for the relay in Figures 1 and 24 uses a d.c. supply of 12 V and a coil current of 0.1 A. It can control a 240 V circuit carrying a current of 10 A.

In Figure 24, find the core of the electromagnet, the coil of the electromagnet, and the two pin connections which lead to the coil (pins A and B).

Find the pivoted arm, or armature, made of magnetically soft steel, and find the moving contact which is fixed to the end of it.

On a sheet of notepaper, draw the relay as in Figure 24(a), and label it. Leave plenty of space below your diagram, for drawing in circuits.

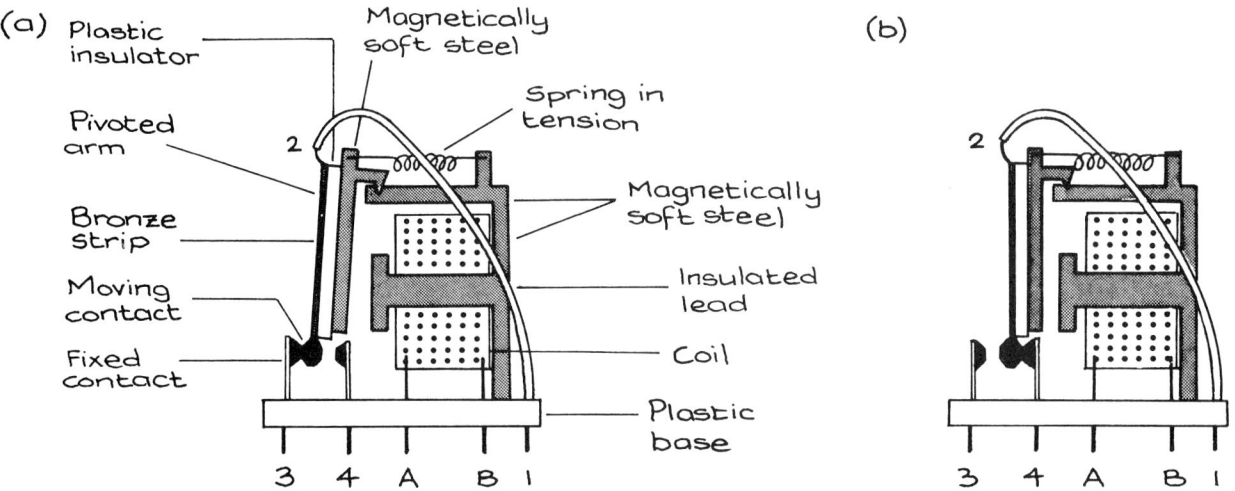

Figure 24 This is a cross-section of the relay in Figure 1. (a) No current through the coil; (b) Current through the coil.

Connecting up a relay

Circuit 1: the control circuit
Add to your diagram of the relay by making a circuit for the coil. Draw and connect to pin A and pin B a 12V battery, and a switch. This is the operator's circuit, or **control circuit.**

Circuit 2: a motor circuit
To the relay, now add a motor circuit. Use pin 1 and pin 4, and draw in a 240V power supply, a symbol for an electric motor, and an ammeter (reading to 10A).

Start up the motor. In your control circuit close the switch. What happens to the electromagnet in the relay? What happens to the pivoted arm, the armature? What happens at contact 4? What happens in the motor circuit?

Stop the motor. In your control circuit, open the switch. What happens to the electromagnet in the relay? What happens to the moving arm, or armature? Does it move? If so, what causes it to move? What happens at fixed contact 4? What happens in the motor circuit? What happens to the motor?

Circuit 3: some improvements

Make some improvements to your control circuit and your motor circuit.

In the control circuit you decide to put a 12V green light, to tell you whether the supply to the electromagnet is on or not. Whereabouts in the circuit should you place the bulb? Should it be in series or in parallel with the coil?

You decide to have a brake for the motor. Use pin 1 and pin 3. Draw in a symbol for an electric brake, near to the motor. Connect pin 3, the brake, a 240V power supply, and pin 1. (You can use a second 240V power socket, or you can use the same one as for the motor.)

When the control circuit switch is **open** is the brake on or off?

When the control circuit switch is **closed** is the brake on? If not, why is it not on?

When the control circuit switch is opened again, what happens to the brake?

An automatically controlled brake

Figure 25 shows the loading trolley for the concrete-making plant in Figure 3 and Figure 4. It raises sand and gravel from ground level up to the mixing pan near the operator's cabin. It is one of the parts of machinery which you may have operated on page 4, operation 8 and operation 11. In Figure 4, the top of the steel runway for the trolley is at 1 o'clock to the engineer's helmet.

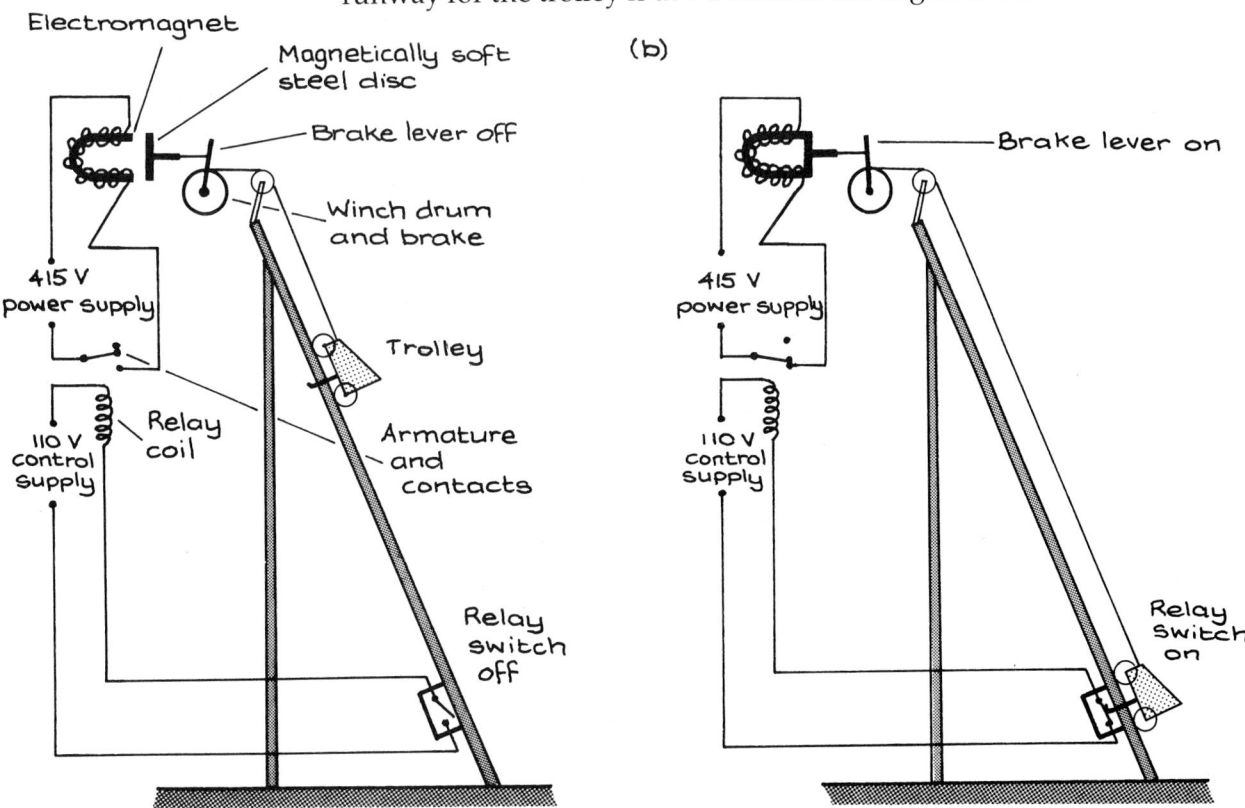

Figure 25 This shows the automatic braking system for a loading trolley. It is the loading trolley for the concrete-making plant in Figure 3 and Figure 4.

In Figure 25(a), the loaded trolley is raised up by a winch drum and cable, driven by an electric motor. There is a brake on the winch drum. When the trolley is empty and is returning to ground level it is still heavy because it is made of steel. It has to be braked to prevent it from crashing into the ground. There is an automatic braking system. From Figure 25(a) and (b) work out how it operates.

On a sheet of notepaper, make a record of the parts and how you think they operate.

1. Find the control circuit. What voltage does it work on? What are the main parts in the circuit?
2. Find the brake-operating circuit. What voltage does it work on? What are the main parts in the circuit?
3. The trolley has a part which sticks out below it (to the left in Figure 25(a)). What does this part do when the trolley nears the bottom of the runway?
4. The trolley is getting near to the bottom of the runway and is approaching the relay switch. Describe, step by step, what happens until the trolley is stopped.
5. A 110V/415V relay is very similar in design to the relay in Figure 24. Sketch Figure 24(a); and draw in the trolley control circuit. Now draw in the power and brake circuit.

Electromagnetic clutches and brakes

When a car stops at traffic lights, the driver keeps the engine running. Figure 26 shows a car stopped at traffic lights. The car is stopped, but not the engine. It would be very inconvenient in crowded towns and cities for a driver to have to switch off the engine every time the vehicle had to stop for pedestrians, traffic jams, and traffic lights. It would also be bad for the self-starting mechanism, for noise, and for fuel consumption. Instead, the driver uses a **clutch**. A clutch will **connect** the engine to the transmission, and will **disconnect** the engine from the transmission.

Figure 26 A vehicle can be stationary and still have its engine running. It uses a clutch.

Figure 27 A clutch transmits a twisting force.
(a) The clutch plates are apart, and no rotation is transmitted.
(b) The clutch plates have been forced into contact. A twisting force and rotation are transmitted. (Adapted from a diagram in *Packaged Starts and Stops*, Warner Electric.)

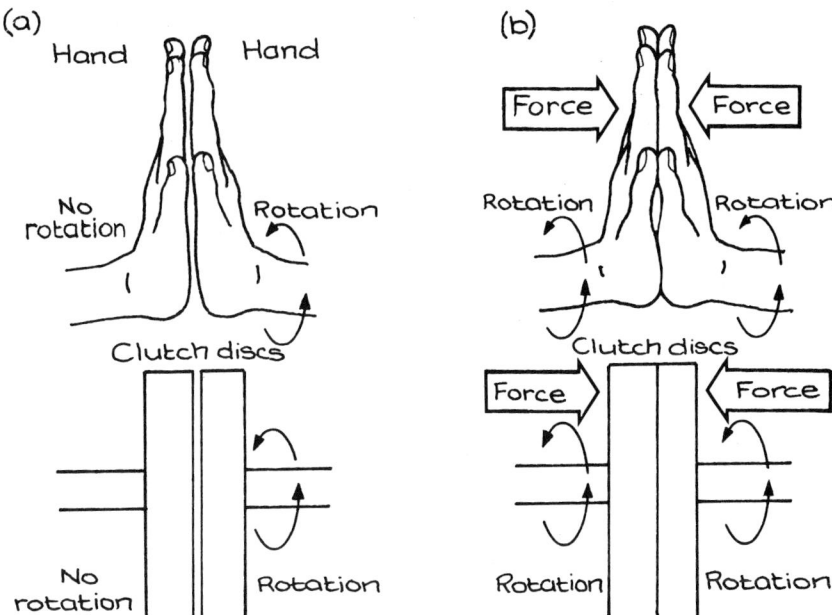

In industry, too, there is machinery that may have to be started and stopped several times a minute. Instead of starting and stopping the engine so frequently, it is often better to use a clutch. Electric motors can only be stopped and started about six times a minute before they overheat. Any faster stopping and starting needs a motor and a clutch.

Figure 27 shows the way in which a rotating motion can be transmitted from one wrist to another wrist. It also shows how a clutch works.

In Figure 27(a) the hands are apart, the rotation of the right hand does not affect the left hand. In Figure 27(b) the hands are being pressed together and the rotating right hand causes the left hand to rotate. The **friction** between the hands transmits a **twisting action** to the left hand.

In Figure 27 the clutch consists of two discs. They have inside faces which have high-friction surfaces. In Figure 27(a) the discs are apart, and the rotation of the right hand one does not affect the left hand one. In Figure 27(b) the discs have been forced together. The friction between the discs transmits a twisting action from the right disc to the left disc; and the left disc and its shaft rotate.

So a clutch is a **power transmission** component. It will transmit power **from an engine to a machine** that is doing work on a load.

In a car, the clutch discs, or plates, are kept forced tightly together by powerful springs. A driver forces the plates apart by pressing down the clutch pedal. In this way the clutch pedal and the clutch are used to connect the engine to the transmission, and to disconnect them.

In very many machines the two plates of a clutch are pulled together and held firmly together by an electromagnet. This is an **electromagnet clutch**.

Figure 28 This shows an electromagnetic clutch.
(a) Electromagnet off; clutch discs apart.
(b) Electromagnet on; clutch discs pulled together. (Adapted from a diagram by Warner Electric.)

Figure 28 shows an electromagnetic clutch. An electromagnet is fixed on the shaft of the machine which has the load. The armature, or piece which will be attracted towards the electromagnet, is on the motor shaft. It is free to move along the shaft, along a groove in the shaft.

Figure 29 shows the groove, in a cross-section of the motor shaft and the armature. On the armature is a projection or **spline**. This fits into the groove. When the motor shaft rotates, what happens to the armature? Why?

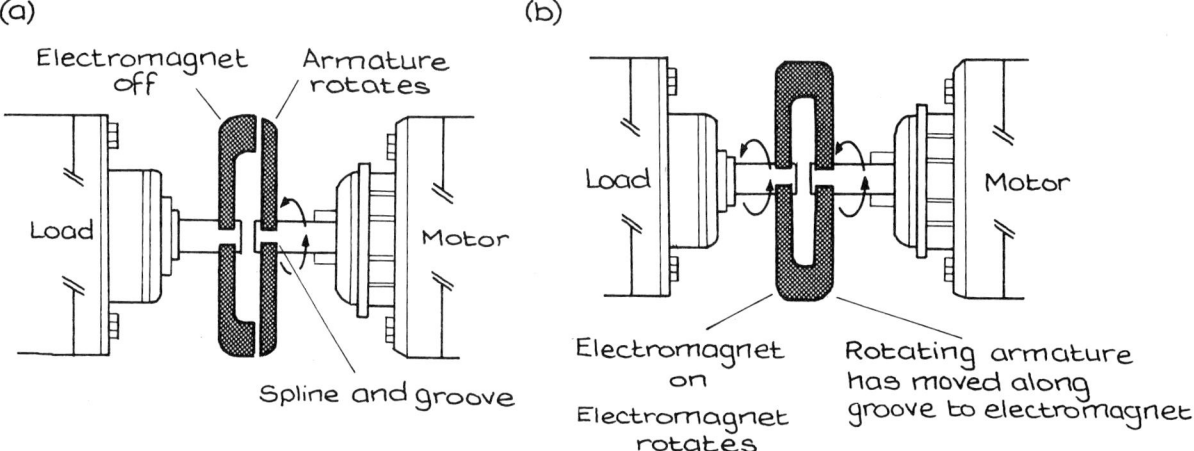

Figure 29 This is a cross-section of a motor shaft and armature. It shows a groove and spline, which make the armature rotate with the shaft.

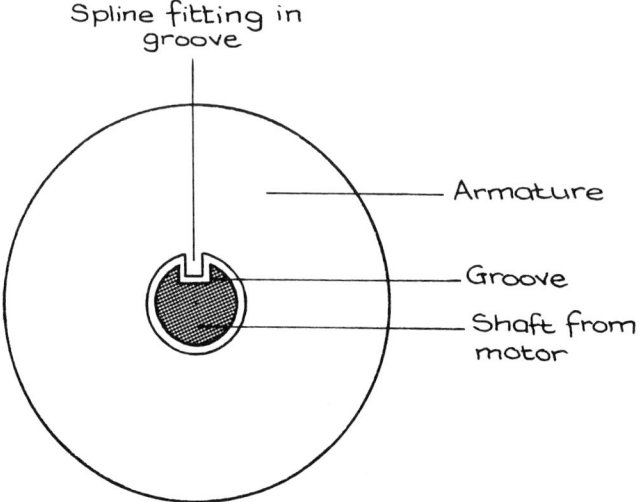

On a sheet of notepaper describe how the system works. You could do it in stages. At the start, the load is at rest, and the motor is stationary.
1 What happens when the motor is started?
2 What happens when the electromagnet is switched on?
3 From your knowledge of electromagnets and armatures, do you think the armature will be made of magnetically hard steel or magnetically soft steel? Why?

4 After the job has been done, the electromagnet is switched off. What happens?

The need for some extra parts to this simple system may occur to you. (What about getting the armature back to where it started from? What about getting current into and out of a rotating electromagnet, without the wires getting wrapped round and round and round the shaft? Have you any ideas for solving these problems? Draw some sketches.)

An electromagnetic brake

When a motor is driving some machinery, the machinery may have to be stopped quickly. The motor may be switched off, but a brake is needed too. Figure 30 shows a motor, some machinery with a load, and an electromagnetic brake.
On a sheet of notepaper show that you know how it works.
1 In the electromagnetic **clutch**, was the electromagnet free to move or was it rigidly fixed? Why was it free to move or rigidly fixed?
2 For the brake, name the parts, A, B and C. (You can look at Figure 28).
3 Is the electromagnet free to move or is it rigidly fixed? Why is it arranged this way?
4 In the arrangement in Figure 30, the motor has to be switched off when the brake is applied. Do you think that this is likely to be a convenient or an inconvenient arrangement? Can you think of a better system?

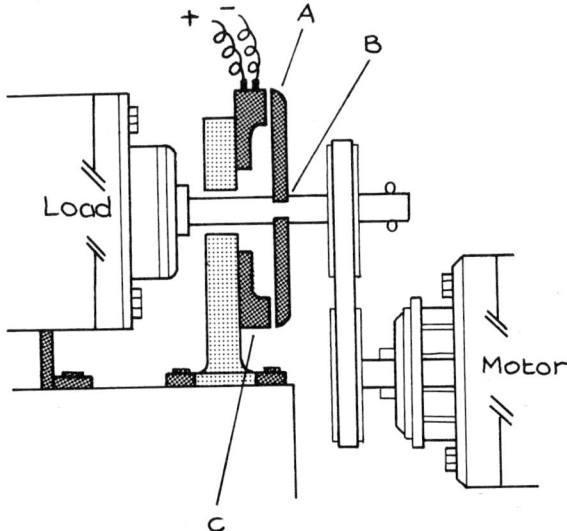

Figure 30 This shows an electromagnetic brake. (Warner Electric.)

An electromagnet clutch–brake assembly

A driver who stops a vehicle puts on the brake, and also presses the clutch pedal. The clutch disengages the engine from the load (the load being the vehicle, its wheels, and transmission), so the engine continues to run, and the vehicle is stopped by the brakes. A vehicle has brakes and a clutch.

Figure 31 This shows an electromagnetic clutch–brake assembly. (Adapted from a diagram of Warner Electric.)

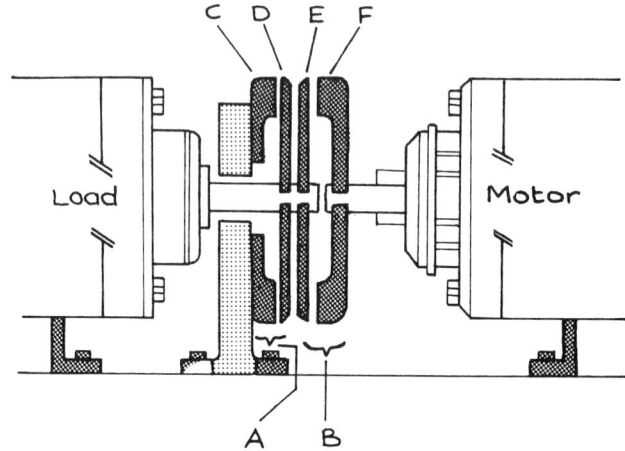

If machinery in industry has to be started and stopped very frequently, it is usually best to have a clutch and a brake. Figure 31 shows a clutch–brake assembly for a motor and its load. Study the Figure. Using a sheet of notepaper to answer the questions, see if you know how the assembly works.
1 Is the motor connected to the load?
2 Which letter shows the whole clutch assembly?
3 Which letter shows the whole brake assembly?
4 Give the letters for
  (a) the clutch electromagnet;
  (b) the brake electromagnet;
  (c) the clutch armature;
  (d) the brake armature.
5 Out of C, D, E and F, which will have a spline and groove?
6 Sketch the clutch assembly when the clutch electromagnet is on.
7 Sketch the clutch–brake assembly when the brake electromagnet is on.

How the electromagnet is wound

When it is seen face-on, the electromagnet is circular. When it is seen in a section through a coil, it is double U-shaped. Figure 32 shows this. The coil is wound in the trough.

The open end of the U has high-friction material placed in it, because this is the face which will be in contact with the armature disc.

Have a look at the poles of the electromagnet in Figure 32. At the top of the circular magnet, find the outer pole and the inner pole.
1 At the bottom of the circular magnet, what is the polarity of the outer pole? What will be the polarity of the outer pole at the top of the circular magnet?
2 At the bottom of the circular magnet, what is the polarity of the inner pole? What will be the polarity of the inner pole at the top of the circular magnet?

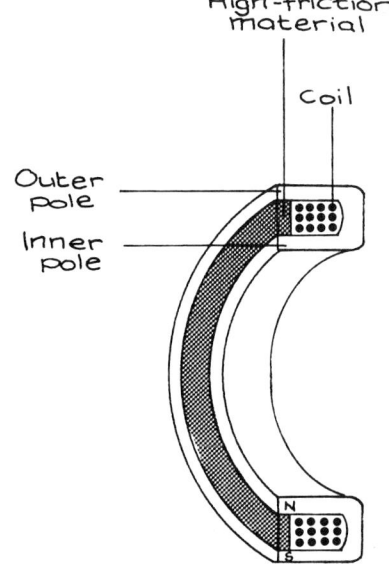

Figure 32 This shows how the coil is wound in the electromagnet. (Warner Electric.)

Figure 33 Brushes and slip rings are used to lead current into a rotating coil and out again. (Adapted from a diagram of Warner Electric.)

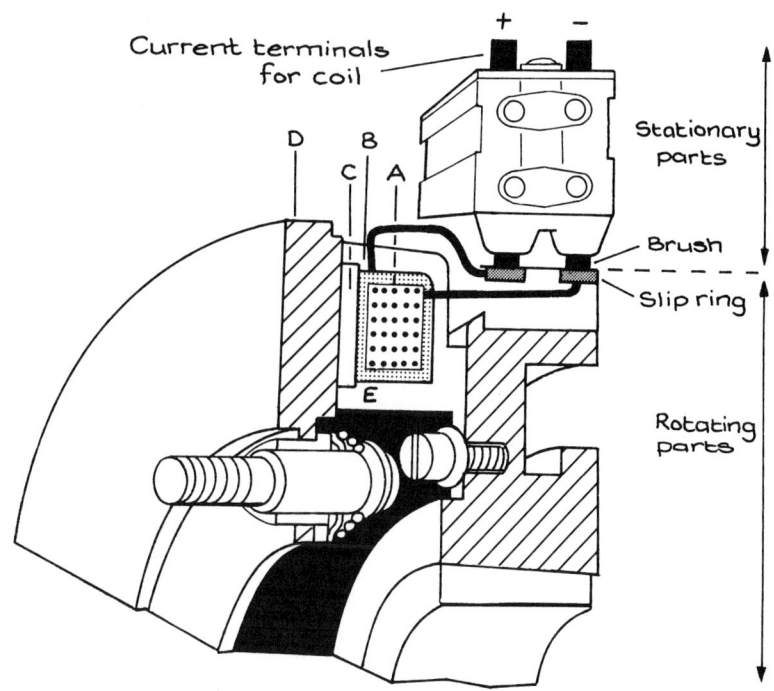

Supplying a rotating coil with current: brushes

In Figure 31, the clutch electromagnet, F, rotates all the time that the motor is turning. How can current be led into the rotating coil and then out again? This is done by two brushes and two slip rings.

Find a slip ring in Figure 33. It is a strip of bronze that goes right round the circumference of the clutch. At one point a lead connects it to the coil.

Find a brush. It rests on a slip ring. It is made of graphite.

The brush is attached to a stationary part of the machinery, and pressed gently onto the slip ring by a spring. The brush is connected to a current terminal for the electric power supply.

There are two slip rings and two brushes. When the clutch rotates, the two slip rings rotate with it. On each slip ring a brush touches lightly, carrying current either into the coil or out of it.

Have a look at the other parts of the clutch.

On a sheet of notepaper, write down the names of the parts A, B, C, D and E. You can choose from armature, outer pole, high-friction material, inner pole, and coil.

Putting the pieces together

Figure 34 shows some very beautiful machinery. It is pleasing to look at. It has been designed with imagination and ingenuity, and made with care and skill. It is produced to do a job reliably at least 5 000 000 times without going wrong.

Figure 34 This is an electromagnetic clutch and brake unit. The photograph is approximately actual size. (Warner Electric.)

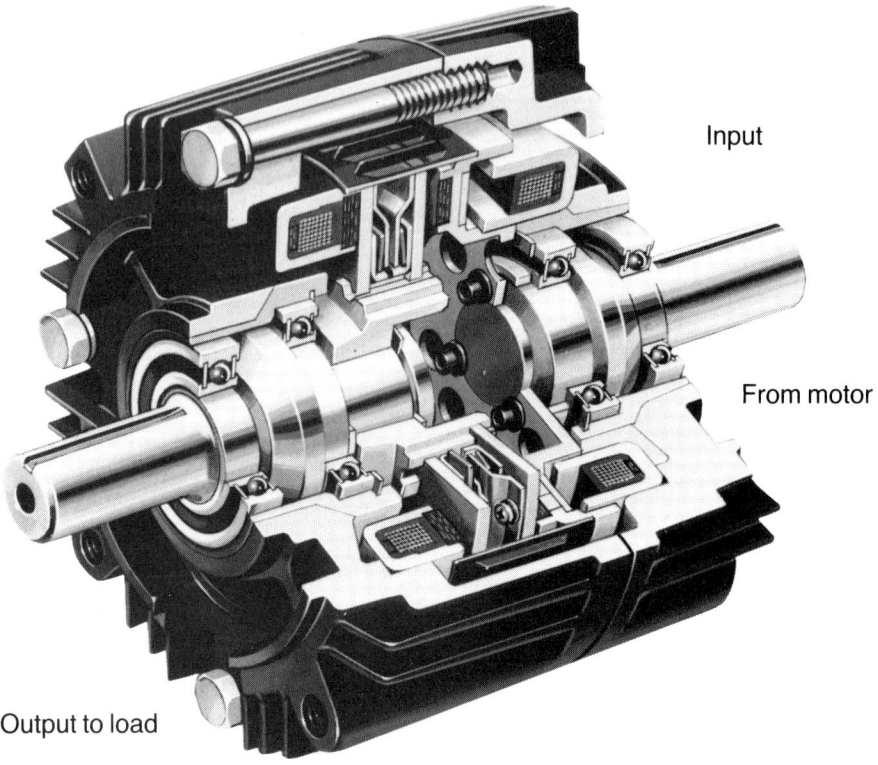

Input

From motor

Output to load

It is a clutch–brake unit. Find how many parts you can identify.
After having a look at Figure 34, look back at Figure 31 and Figure 32 to remind yourself what parts to look for.

On Figure 34 see if you can find the main parts, and use a sheet of notepaper to answer the questions.
1 Find the clutch coil.
2 Find the brake coil.
3 Find the U-shaped magnet, for the clutch.
4 Find the U-shaped magnet, for the brake.
5 Find the high-friction material for the brake.
6 Find a groove and spline for the brake armature.
7 Find the brake armature. (This is very clear in the bottom part of the photograph.)
8 Is there just one groove and one spline for the brake, or are there many grooves and splines, all the way round the circumference?
9 Find the clutch armature.
10 For the clutch armature are there one, few, or many grooves and splines?
11 The input shaft must rotate smoothly in the unit. Find two sets of ballbearings.
12 Do the same for the output shaft.
13 If you wished to take the unit to pieces where would you start?
14 All the way round the unit there are metal fins. What are these for?
15 Where does the heat come from?

The following are some sources. List them in order, with the greatest source of heat production first: friction at the ball bearings; heat from current in the clutch coil (the clutch coil uses less than 40 watts of power when used with a 30 kilowatt motor); heat from current in the brake coil (the brake coil uses less than 40 watts of power when used with a 30 kilowatt motor); friction when the clutch is engaged (the clutch surfaces will slip on each other until the output shaft is rotating as fast as the input shaft); friction when the brake is put on (the brake surfaces will slip on each other until the output shaft has come to rest); other sources of heat.

Jobs in industry for clutches and brakes

Figure 35 shows part of a spinning frame. If a thread breaks, the machinery is instantly stopped by an electro-magnetic brake.

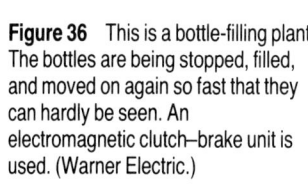

Figure 35 This is part of a spinning frame. It uses an electromagnetic brake to stop the machinery if a thread breaks. (Warner Electric.)

Figure 36 shows bottles being filled. The bottles are moved on a conveyor system. The conveyer has to be stopped so that each bottle is under a delivery nozzle, for filling. The conveyor then

Figure 36 This is a bottle-filling plant. The bottles are being stopped, filled, and moved on again so fast that they can hardly be seen. An electromagnetic clutch–brake unit is used. (Warner Electric.)

Figure 37 These are two graphs for a machine in a production plant. They show how the machine speed changes with time.
(a) The machine is connected directly to the motor.
(b) The machine is connected through a clutch–brake unit to the motor.

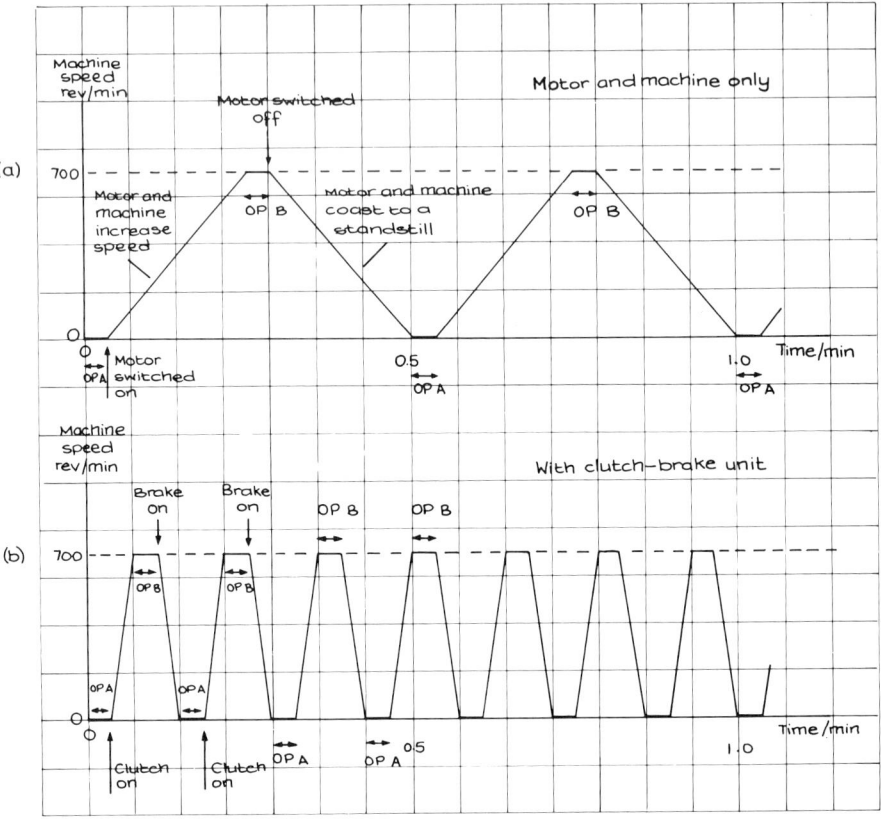

has to be started again. The stopping and starting are done by an electromagnetic clutch–brake unit.

The bottles must be stopped in exactly the right place, so that they are exactly under the delivery nozzle. This can be done automatically because the clutch and brake are electric, and they can be switched on and off automatically by a sensing device.

Figure 37 shows two graphs for a machine which is part of a production process. When the plant was set up, the machine was driven directly by an electric motor. Two operations had to be done, A and B. Operation A (OP A) had to be done when the machine was motionless; and operation B (OP B) had to be done when the machine was rotating at a speed of 700 revolutions per minute.

Graph (a) shows how the speed of the machine changed with time, in the days when the machine was driven directly by the electric motor. At the start, the machine is motionless (machine speed = 0 rev/min) and Operation A, OP A, is done. Then the motor is switched on, and the motor and machine increase speed until a speed of 700 rev/min is reached. At that moment, the speed becomes steady and Operation B, OP B, is done.

Then the motor is switched off, and the motor and machine slow down, coasting until they come to rest. That is one complete cycle. Then the cycle starts again.

How many cycles were completed in 1 min?

Was most of the time spent doing the important work of operations A and B, or was most of the time spent starting and stopping?

Electric motors connected to a load start up rather slowly, and take rather a long time to reach their design running speed. Then, when they are switched off they take rather a long time to slow down and stop.

Production was slow in the works because of this. The engineers and technicians discussed the problem and decided to use a clutch and brake unit between the motor and the machine. The arrangement would then be like the one in Figure 31; have a look at this again.

The motor would be kept running continuously, at its design speed. The machine would be started and kept running by using the clutch; and it would be stopped by using the brake, and disengaging the clutch. Figure 37(b) shows the result.

Have a look at Figure 37(b). Work from the start and follow through the first cycle.

Now follow through the second cycle.

How many cycles does this new arrangment complete in 1 min?

How many cycles did the old system complete in 1 min?

How many times greater is the productivity?

Very high speed equipment is needed in the packaging industry. In a bag-making plant, a well designed clutch and brake system can give starts and stops at 300 cycles per min. How many cycles per second is this?

In the printing industry, parts of the machinery have to start and stop at rates of many cycles per min. In the textile industry the same is true.

In these three industries the machinery uses electromagnetic clutches and brakes, because these can give very high cycle rates.

1. On a sheet of notepaper, write a summary of the reasons why electromagnetic clutches and brakes can do very many useful operations in industry.
2. The engineer in charge of the plant that gave the graph in Figure 37(b) was still not quite satisfied. The machine was now being shaken too much by the rapid starts. He wanted the clutch to go on more gently. He wanted the machine to take longer to get up to 700 rev/min.

Suggest a good way in which he could do this. How could the working of an electromagnetic clutch be altered without touching the clutch parts at all?

Draw a very simple scheme by which you could vary the action of the clutch, from fierce and grabbing to gently and slipping. (**Hint.** You may wish to use a variable resistor.)

Mark your diagram so as to show how to get fierce, medium and gentle starts.

Making alternating current electricity

What are alternating current generators?

The electricity for industry and for our homes is made by electromagnetic induction. A magnetic field is made to rotate inside a stationary coil of wire, and an electric current can be taken from the coil. The dynamo on a pedal cycle uses this method, and so does a power station on the electricity grid. They both supply a current which reverses its direction through the circuit many times per second. This is why it is called **alternating current,** or reversing current.

The dynamo on a pedal cycle supplies electricity at a potential difference of 6 volts, and it will supply about 4 watts of power. The source of this power is the person who is pedalling the cycle.

The generator in Figure 38 can provide electricity at 220 V or 110 V, and it will supply 5000 W, or 5 kilowatts of power. This is more than enough power for the 1.5 kW circular saw which is lying on the carpentry bench behind it. In fact the generator would provide the power for two 1.5 kW circular saws and two 1 kW floodlights all in use at the same time.

A generator has two main parts: the **engine** and the **alternator** or dynamo. The engine in Figure 38 is a petrol engine of about 6 kW output, and it is on the righthand side. It turns the alternator. The alternator is on the left, from the chassis up to the guard rail. The rectangular box above it contains the sockets for plugs, and the voltage selector (220 V or 110 V).

In a large power station, a steam turbine usually drives the alternator. The output of the alternator could be 50 000 000 W, that is, 50 000 kW or 50 megawatts of power; and it would usually be at 11 000 V. The source of heat for making the steam is usually the burning of a fuel: for instance air and oil, or air and coal. The heat could also come from a nuclear reactor.

Figure 38 This shows a 5kW generator of alternating current. It can be pushed easily from place to place by one person. Above the wheel is a petrol engine. This turns the alternator, or dynamo, which is to the left of the wheel.

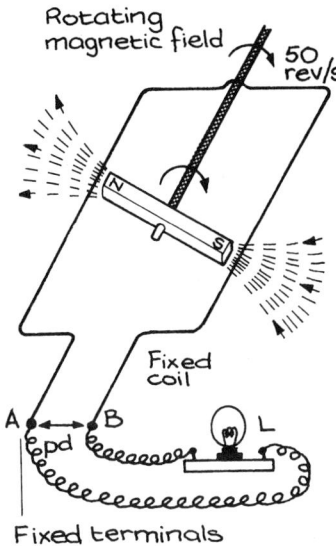

Figure 39 The magnetic flux which is rotating in the coil of wire produces a current in the circuit.

In Figure 39, a bar-shaped permanent magnet is being rotated inside a fixed coil of wire, and the coil is connected in a circuit to a lamp. In the diagram, magnetic flux is threading through or linking through the coil. As the magnet is rotated, the flux which is threading through the coil changes. This creates, or induces, an electric current in the circuit.

Follow the rotation of the bar magnet. In the diagram, the N pole is pointing up through the coil. Half a turn later the N pole will be pointing down, and the S pole will be pointing up. The direction of the magnetic flux through the coil will have reversed, and the current will have reversed

The graph in Figure 40 shows this happening. The current going in the **forward direction** in the circuit grows, and then it decreases to zero. The current then reverses, and it starts growing in the reverse direction. The size of the reverse current then decreases to zero. All of this happens in one complete revolution of the magnet, that is, one cycle.

If the bar magnet is rotated at 50 revolutions per second, 50 rev/s, then the current does 50 cycles per second. This is a frequency of 50 hertz. The 50 cycles include 50 reversals in each second. How long does it take for one cycle to happen? Check by looking at Figure 40.

The electricity which is supplied to our homes from the grid is alternating current with 50 cycles per second, 50 Hz. The portable generator in Figure 38 also provides alternating current with a frequency of 50 Hz.

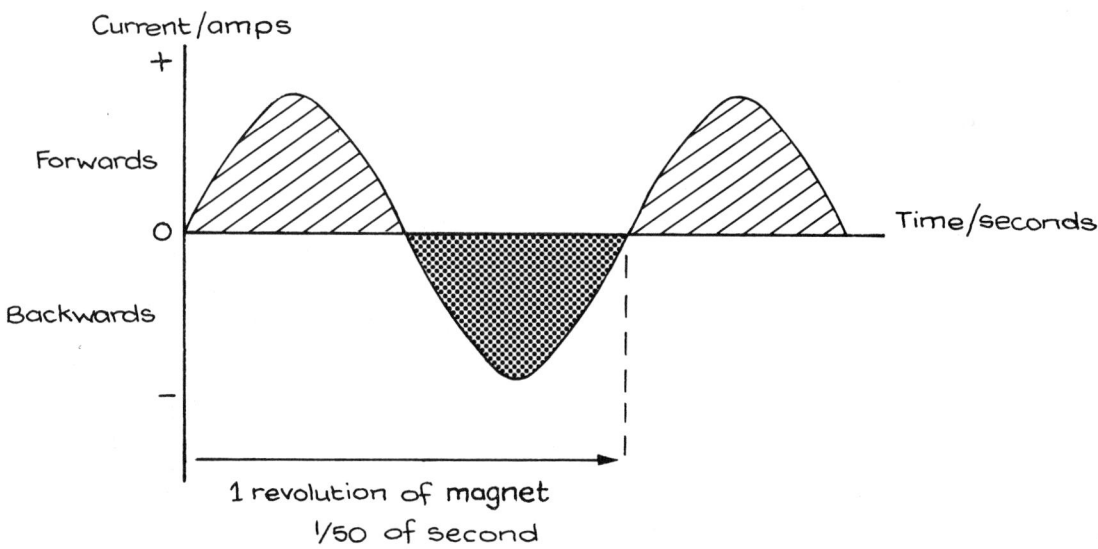

Figure 40 This graph shows how the current in the circuit of Figure 39 changes with time. The current grows and falls, and then reverses. It is alternating current.

The rotor

In most alternators the rotating magnetic flux is produced by an electromagnet. This is called the rotor. Figure 41 is a drawing which shows the main parts of a rotor. Have a look at the parts. Use a rule that you know, and check that the North pole and the South pole have been marked correctly.

The petrol engine will drive the shaft and the rotor round at 3000 rev/min. How many rev/s is this?

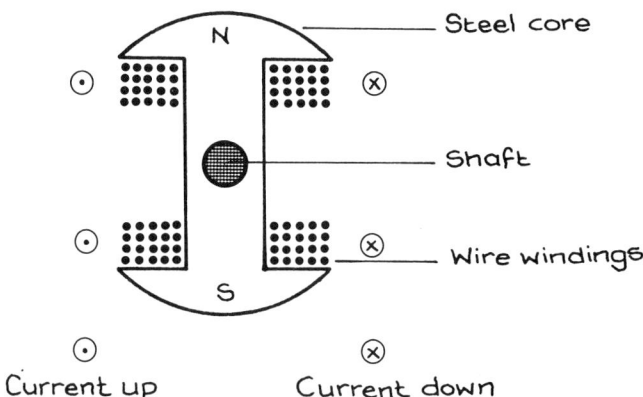

Figure 41 This is a cross-section of an electromagnet for producing a rotating magnetic field. It is called a rotor.

In the generator in Figure 38, the rotor has 500 turns of copper wire on the core. The wire has a diameter of 1 millimetre, and it is insulated with varnish. When the machine is working, the current through the windings is about 5 amps.

If you tie a stone onto the end of a piece of string, you can whirl the stone round and round your head. But you can only do this as long as you keep pulling inwards on the string. If you let go of the string, the stone flies away. It is the same with the copper wire on the rotor. It rotates at 3000 rev/min, and the wire has to be held in firmly to prevent it from flying off the rotor and destroying the machine.

The copper wire of the rotor has a mass of about 1 kilogram, and it is held in two ways. Have a look at Figure 41. This shows that the wire is wound **under** the ends of the steel poles of the electromagnet; and these poles hold the wire in place, up and down. Figure 42 shows three bands going horizontally over the rotor, from pole to pole. These are steel strips, and they hold the wire in place.

The inwards pulling force which is needed to keep the wire in place at 3000 rev/min is equal to the weight of a ½ tonne mass. Designing and making the rotor to allow for this needs careful calculation and good workmanship.

The current for the coils has to be led into and out of the electromagnet while it is rotating on the shaft. This is done by two **slip rings.** These can be seen in Figure 42 just to the left of the

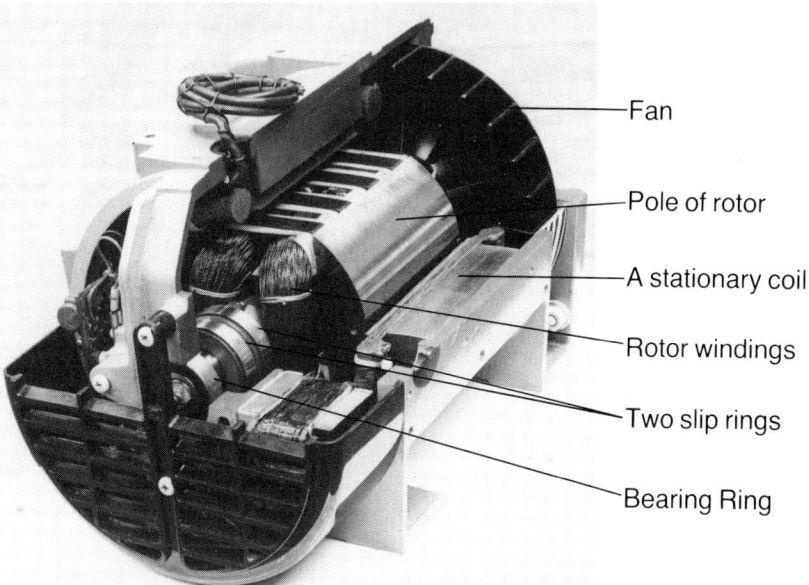

Figure 42 This photograph shows the assembled alternator with part of it cut away. See how many parts you can recognize. (Markon Engineering Limited, UK.)

coil. When the machine is assembled, a short carbon rod presses lightly onto one slip ring and another rod onto the other slip ring. These rods are called **brushes** and they make a slipping contact through which current can reach and leave the slip rings.

On the right of the rotor, in Figure 42, is a structure with many flat blades. This is a **fan**. It is for drawing air through the alternator to cool it. The coils get very hot; their operating temperature is about 120 °C. Why do the coils get hot? Where does the heat come from? Could it come from air friction on the coils rotating at 3000 rev/min? From the petrol engine? From the 5 A current in the coils?

On the extreme left of the machine in Figure 42 is a third ring. This is a **bearing ring**. Between it and the shaft there are ballbearings. The bearing ring fits into the main frame and supports the end of the rotating shaft.

The stationary coil

The section in Figure 42 cuts through a **stationary coil.** It is on the righthand side of the machine. It is quite close to the curved pole of the rotor.

The diagram in Figure 39 shows one turn of wire in the fixed coil. A real alternator has several turns of wire in the fixed coil. Our 5 kW alternator has twenty turns, made of copper wire insulated with varnish.

The working alternator gets hot

The description of the rotor said that it used a current of about 5A, in its copper wire winding. Now, a current in a conductor raises the temperature of the conductors, so the rotor winding, and the steel core, become hot.

The resistance of the hot, working rotor winding is 14 ohms; and the current in it is 5 A. From this information we can work out how many joules of heat are produced per second by the current in the rotor winding.

If I = current in the conductor, and R = resistance of the conductor, then

$$\begin{aligned}\text{Rate at which heat is produced} &= I^2 R \\ &= 5 \times 5 \times 14 \quad \text{J/s} \\ &= 5 \times 70 \quad \text{J/s} \\ &= 350 \, \text{J/s}\end{aligned}$$

So the current in the rotor winding produces heat at the rate of 350 J/s, that is, 350 W.

When the alternator is working at full power, heat is produced in the stationary coil at a rate of 700 J/s, that is, 700 W. Therefore, the total heat production from the rotor coil and the stationary coil is 700 W + 350 W = 1050 W. This is just over 1 kW, and is equivalent to a powerful hair-dryer or a domestic convector heater of low power.

If the inside of the alternator were not cooled in some way, the temperature of the coils would rise higher and higher. The insulation would become destroyed, and the coils would become destroyed. A fan is used to draw a stream of cooling air over the coils.

The fan is mounted on the shaft, and it rotates at the shaft speed of 3000 rev/min. Have a look for the fan in Figure 42. The fan draws about 2 metre^3 of air per minute over the windings. Even with this cooling, the temperature rise of the coils is about 100 K. So, if the atmospheric temperature were 20 °C, the working temperature of the coils would be about 120 °C, or well above the boiling temperature of water.

Efficiencies of alternators

Our alternator receives power along the shaft from the engine, and it converts most of this power into electrical power. But about 1 kW is lost.

Energy conversion always leads to some loss of **useful** energy. None of the original energy is actually destroyed, but some of the forms into which it changes may not be useable. Heat and sound may be produced, and it may not be possible to use these.

The efficiency of a machine is defined as:

$$\text{Efficiency} = \frac{\text{useful energy got out}}{\text{total energy put in}} = \frac{\text{useful power got out}}{\text{total power put in}}$$

The manufacturer of the alternator in Figure 38 gives this data about it. Maximum power received by alternator from the shaft = 6.2 kW; maximum electrical output = 5.0 kW.

$$\text{Efficiency} = \frac{\text{useful power got out}}{\text{total power put in}} = \frac{5.0 \, \text{kW}}{6.2 \, \text{kW}} = 0.81 = 81\%$$

The alternator is therefore 81% efficient.

The firm makes several models with the same basic design but different output power. Table 1 gives the information.

Table 1 Power input and power output for different models of alternator. (Markon Engineering Limited, UK.)

Model	Input from shaft / kW	Electrical output / kW	Efficiency / %
A	3	2	
B	4	3	
C	5	4	
D	6.2	5	81

With most electrical machinery, the higher the full output, the higher the efficiency. An alternator with 100 times the output of our 5 kW machine, that is, with an output of 500 kW, has an eficiency of about 93%. A very large turbogenerator in a national grid power station can have an output 1000 times larger still, that is 500 000 kW, or 500 MW. The efficiency of its alternator can be about 99%. It has to have its stationary coils cooled by water, and its rotor coils cooled by a stream of hydrogen gas.

A very large generator has the advantage of very high efficiency, but it cannot be moved. A small generator has the advantage that it can be moved easily from job to job, but it has a lower efficiency. For very many jobs, a small portable generator is needed. For these alternators, efficiencies of 70% to 80% are very acceptable; and they are the results of good design and good workmanship by the makers.

Putting the parts together

Figure 43 shows parts of the 5 kW alternator taken apart. How many parts can you recognize and name? On a sheet of notepaper, give each numbered part its name.

You can choose from: two brushes, two slip rings; fan; grille to let in cool air; wire cage to let out hot air; pole of rotor; windings of rotor; steel bands; bearing rings; and fixed coils (for effectiveness, the machine has more than one fixed coil).

Parts 3 and 7 are a long way apart on the diagram. Do they have anything to do with each other in their functions? If so, how do they come together?

Are the parts in 2 essential when the machine is stationary? Are they essential when the machine is rotating at 3000 rev/min? If so, describe briefly what they do, and why they are essential. Does 9 help at all with this mechanical need?

Figure 43 This is a diagram of the main parts of the alternator. (Markon Engineering Limited, UK.)

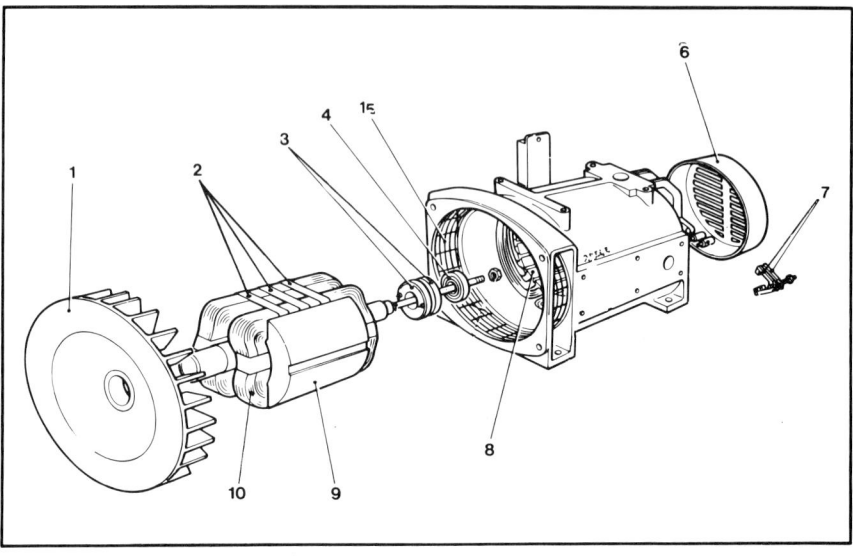

A large alternator

Figures 44 and 45 show a large alternator being made. It will produce 6 MW of power (6000 kW) at 11 kW (11 000 V). It could be used to provide power for an isolated community, or it could feed power into an electricity grid.

In Figure 39, page 33, the simple alternator had one fixed coil, and one pair of poles on the rotor. An alternator is more effective if it has several fixed coils, and several pairs of poles on the rotor.

Figure 44 shows that there are many fixed coils. Figure 45 shows that there are several poles on the rotor: how many poles, and how many pairs?

Figure 44 The photograph shows the stationary coils of a 6000 kW (6 MW) alternator. The alternator produces power at 11 000 V (11 kV). (Brush Electrical Machines Limited, UK.)

Figure 45 This is the rotor for Figure 44. How many poles does the rotor have? How many pairs is this? (Brush Electrical Machines Limited, UK.)

Transformers

What are transformers for?

Tranformers readily **change the voltage** of an electricity supply. Tranformers can change the voltage **up**, and they can change it **down**.

A large power station which is supplying the electricity grid in the United Kingdom will generate electricity at 25 000 volts (25 kilovolts). A transformer steps this up to 400 000 V (400 kV) and then feeds it into the grid cables. The reason for doing this is connected with efficiency and cost. It is more efficient and cheap to transfer electricity at a high voltage than at a low voltage.

In our homes and in production works we cannot use electricity at 400 000 V, so a succession of transformers is used to reduce the voltage down to 240 V. The changes down will be from 400 000 V to 132 000 V; then from 132 000 V to 33 000 V; then from 33 000 V to 11 000 V; then from 11 000 V to 240 V. How many transformers may there be between the 240 V supply in your home and a 25 000 V power station where electricity is produced?

The transformer in Figure 46 takes in power at 415 V and feeds it out at the safer 110 V. In the photograph, it is providing power for the 110 V circular saw on the trestle table.

On a construction site, electricity may be needed at many different voltages: 415 V for tower cranes, 240 V for lighting and heating in offices, 110 V for hand-tools, and 50 V or 25 V for hand-lamps in damp, dangerous places such as tunnels. It would not be economical to have an electricity generator for each of these voltages. The best way is to have a 415 V supply on the site and then to use the transformers.

Figure 46 This is a 10 kW transformer. It takes in power at 415 V and gives out power at 110 V. It is being used to drive the circular saw on the trestle table.

How do transformers work?

Transformers work by **electromagnetic induction**.

Figure 47 shows a very simple transformer. It consists of a **core of magnetically soft steel**. Onto this is wound an inner coil of insulated wire, the **primary coil**. Around the primary coil is wound an outer coil of insulated wire, the **secondary coil**.

Alternating current is supplied to the primary coil, through terminals A and B. The current in the coil produces a magnetic flux through the coil. This flux threads through, or links through, the secondary coil. The secondary coil shares the magnetic flux produced in the primary coil.

As the alternating current in the primary coil grows and falls and reverses, the flux grows and falls and reverses. This shared **changing magnetic flux** creates, or induces, an **e.m.f. in the secondary coil**. In our circuit, the secondary coil is connected to a lamp; and the e.m.f. in the secondary coil produces a current in the lamp circuit.

Think in stages about what happens to the magnetic flux through the secondary coil. Do you think that the current in the secondary circuit is (a) steady direct current, (b) varying direct current, or (c) alternating current?

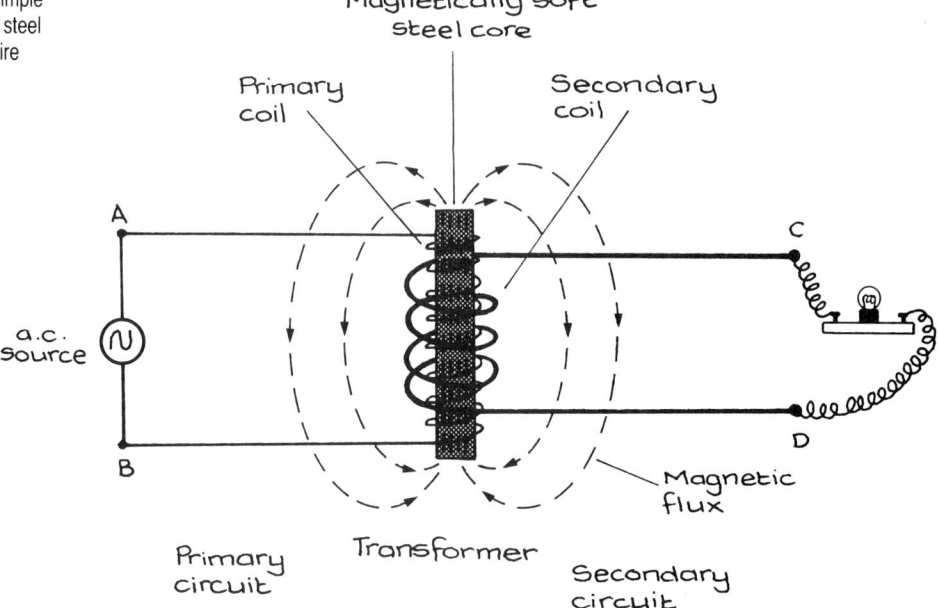

Figure 47 This shows a very simple transformer. A magnetically soft steel core has two separate coils of wire wound on it.

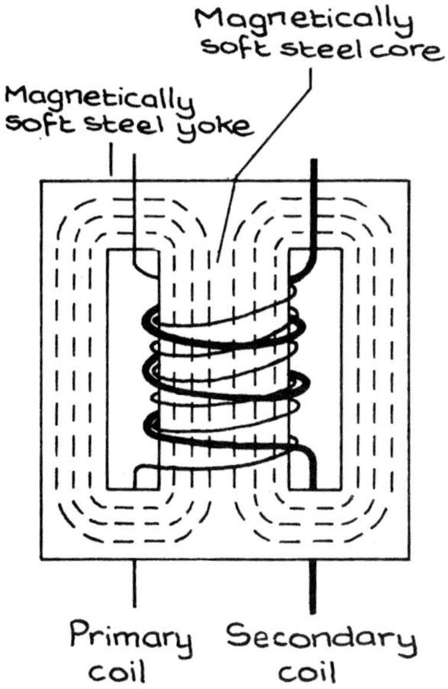

A practical transformer

The arrangement in Figure 47 does work but it is not satisfactory. The magnetic flux is mostly in air, and it is difficult to form flux in air. It is **very much easier to form flux in steel.**

Figure 48 shows a design which has a magnetically soft steel core surrounded by soft steel yokes. When flux forms, **the flux loops can be entirely in soft steel.** This makes a much more efficient machine than the arrangement in Figure 48.

Although the transformer in Figure 49 is quite small (the spectacle case beside it is 15 centimetres long), it will transform 10 kilowatts of power. It is designed to change a supply voltage from 240 V down to 110 V, for electric hand-tools. The efficiency of the transformer is about 96%.

Figure 48 In this transformer the magnetic flux loops are entirely in soft steel. It is a much more effective arrangement than the one in Figure 47.

Power through space

Have a look at Figure 48 again. The primary circuit is separate from the secondary circuit. The wires of the coils are insulated; and in most transformers there is a layer of insulator between the primary coil and the secondary coil. Yet power can be transferred from the primary coil to the entirely separate secondary coil. With the transformer in Figure 49 this can happen at the rate of 10 000 joules per second (10 kW). The power is being transferred through space, with no material connection. This is very remarkable. It is done by the changing magnetic flux.

A changing magnetic flux can be used to transfer power.

Figure 49 This transformer is designed as in the diagram in Figure 48. The thick black part near the spectacle case is a steel yoke. The coil that is sticking out is the secondary coil.

Getting the right voltage change

Engineers need to be able to design and build transformers that will bring about particular voltage changes, for instance 415 V to 110 V. The voltage change that takes place depends upon the ratio of the number of turns in the primary coil to the number of turns in the secondary coil. This is shown in Figure 50.

The transformer in Figure 50 has **fewer turns in its secondary coil** than in its primary coil. It will step the voltage **down**. It has 100 turns in its primary coil and 50 turns in its secondary coil. These are in the ratio of 100:50, or 2:1. The ratio of primary voltage to the secondary voltage to the secondary voltage will also be 2:1.

Figure 50(b) shows that an input voltage of 20 V at the primary terminals. The input voltage:output voltage ratio will be 2:1, or 20:10, so the output voltage will be 10 V.

In the transformer in Figure 50:
1. If the input voltage is 40 V, what will the output voltage be?
2. If the input voltage is 10 V, what will the output voltage be?
3. If the output voltage is 400 V, what must the input voltage be?

The ratio rule gives a very useful equation:

$$\frac{\text{Number of primary turns}}{\text{Number of secondary turns}} = \frac{\text{Primary volts}}{\text{Secondary volts}}$$

$$\frac{N_P}{N_s} = \frac{V_P}{V_s}$$

Figure 50 This is a 2:1 step-down transformer.

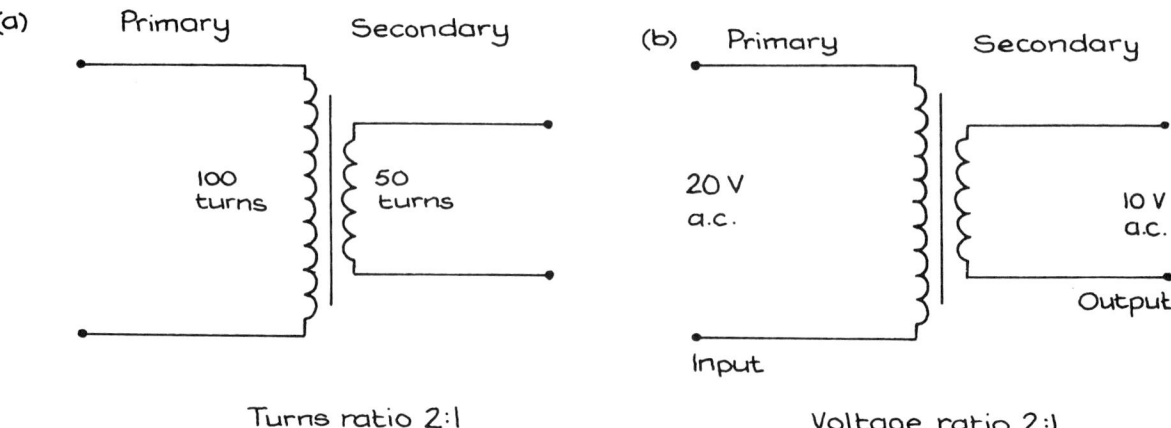

Figure 51 shows a step-up transformer. There are **more turns in the secondary coil** than in the primary coil, and the voltage is stepped up.

The turns ratio is 100:300, or 1:3, so the output voltage is three times the input voltage. If the input voltage at the primary terminals is 30 V, then the output voltage at the secondary terminals is 90 V.

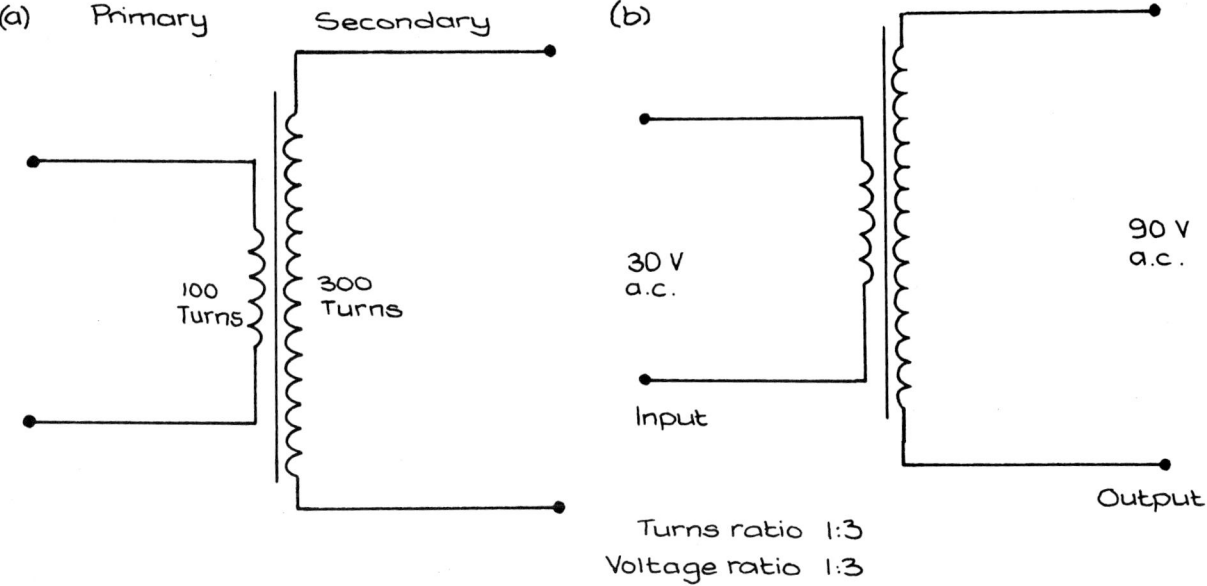

Figure 51 This is a 1:3 step-up transformer.

Figure 52 The technician is completing a 2kW step-down transformer, 240V/110V. She is putting the last part of the core and yoke into place. (Loheat Limited, UK.)

In the transformer in Figure 51:
1. If the input voltage is 10 V, what is the output voltage?
2. If the output voltage is 600 V, what is the input voltage?

The useful equation

$$\frac{N_P}{N_s} = \frac{V_P}{V_s}$$

works for both step-up transformers and step-down transformers. Let us see how it can be used. The company which manufactures the 10kW step-down transformer, 415V to 110V, in Figure 49 makes a range of transformers. They are for different voltage changes, and different power outputs. Have a look at Figure 49.

The company makes a 2 kW step-down transformer, 240 V to 110 V. Figure 52 shows a technician completing work on this. Calculations show that the secondary coil should have 55 turns in it. How many turns should the assembly technician put on the primary coil?

$$\frac{N_P}{N_s} = \frac{V_P}{V_s}$$

$$\frac{N_P}{55} = \frac{240}{110}$$

$$N_P = \frac{240}{110} \times 55 = \frac{240}{2} = 120$$

So the primary coil should have 120 turns on it.

43

How efficient are transformers?

Transformers are one of the most efficient types of machines. There are no moving parts so there are no losses of energy due to friction.

A good diesel engine has an efficiency of about 35%. The small transformers in Figure 46 and Figure 49 have efficiencies of about 96%. A large transformer in the electricity grid has an efficiency of about 99%.

If a transformer is 96% efficient, then 4% of the electrical power that goes in at the primary coil is lost. It is almost entirely lost because the coils become hot. If it is a 10 kW transformer, then 4% of 10 000 W is lost. That is, 4/100 × 10 000 W, or 400 W. This is about equal to the heat given out by a low-powered hair-dryer.

Transformers get hot

An electric current in a conductor raises the temperature of the conductor, so the windings of a working conductor become hot.

Have a look again at Figure 49. This shows the yoke and windings of a 10 kW transformer which is about 96% efficient. This means that 4% of 10 000 W is given out, that is, 400 W, or 400 joules per second.

If the energy cannot escape quickly enough, the temperature of the windings will become too high, the insulation will be destroyed, and the windings will burn out. When the transformer in Figure 49 is working steadily its steel core and inner copper windings will be at a temperature of about 120 °C, that is, well above the boiling temperature of water. The outer part of the yoke will be at about 100 °C.

When an electrical engineer starts to design a transformer that the firm has been asked to make, the very first calculations are to work out how much energy will be produced each second in the coils. The engineer then produces a design that will allow the energy to escape fast enough for the transformer not to overheat. In small transformers, most of the energy is radiated away. This is so for the machines in Figures 46 and 49, of about 10 kW power.

In large machines such as electricity board transformers, the core and the coils have to be cooled by oil. The oil then circulates through pipes or fins in contact with the outside air. Energy is lost to the atmosphere, and the cooled oil recirculates through the transformer. Look for an Electricity Board transformer, and look for the cooling pipes or fins.

Humming transformers. What note? How loud?

Transformers give out a humming noise. Stand by the fence around an electricity board transformer and listen for sound from it.

Try and hum the note which is coming from the transformer. Is it low-pitched, middle-pitched, or high-pitched? Try and find the note on a piano. You will find that it is two Gs below Middle C. It has a frequency of 100 hertz, or 100 cycles per second (in the United Kingdom).

When unmagnetized steel is magnetized its volume is very slightly changed. When steel is magnetized by alternating current it is magnetized **twice** in every cycle; that is, once by the forwards current, and once by the backwards current. If the frequency of the alternating current is 50 Hz, then the steel is magnetized 2×50 times per second = 100 times per second.

So the magnetically soft steel in a transformer core in the UK is magnetized 100 times a second, and **its volume changes 100 times a second. This causes it to vibrate** and to give out a note of frequency 100 Hz.

In the USA, the mains frequency is 60 Hz. Transformers in the USA hum with a note of what frequency? You could find it on a piano; it is two B's below Middle C.

A badly designed and badly made power transformer could give out a lot of noise, and could be a disturbance to people who live near it. Manufacturers of power transformers have carefully studied how noise is generated in the machines and how it is transmitted. The knowledge which they have gained in these studies helps them to design transformers. They can design and build transformers which provide a great deal of power and make very little noise.

Manufacturers of transformers test each transformer when it has been made. They test them for electrical quality and also for sound output. Table 2 shows the results of sound tests on a range of large power transformers.

Table 2   This table is for two different designs for 11000/415 V transformers, Type C and Type T. Each type can be built in different sizes, to suit the different power needs of users. The table shows the noise level at 0.3 m from the transformer, for different sizes (maximum powers) of transformer. (GEC Distribution Transformers Limited.)

Transformer size maximum power/kW	Sound level/decibel		Transformer size maximum power/kW	Sound level/decibel	
	Type C	Type T		Type C	Type T
200	—	51	1600	—	59
300	—	53	1750	60	—
500	55	55	2000	61	61
1000	57	57	2500	—	61
1250	58	58	3000	—	62
1500	59	—	4000	—	64

Draw graphs for the two different types of transformer, as follows.

1 **Type C transformers** are manufactured in several different sizes (with different maximum powers). For Type C transformers

plot a graph of Maximum Power, on the x axis, against Sound Level, on the y axis. Draw the best fit line. Would you say that the Noise Level is proportional to the Maximum Power (Yes/No)?

2 **Type T transformers** are manufactured in several different sizes. Use a **fresh sheet** of graph paper. For Type T transformers plot a graph of maximum power against sound level. Draw the best fit curve. Would you say that the noise level is proportional to the maximum power?

Some more data and questions

Table 3   Some data for a series of air-cooled transformers: 11000V/415V. (GEC Distribution Transformers Limited.)

Maximum power/kW	Power loss (at maximum power)/kW	Mass (of transformer) kg
800	9	3100
1000	11	3400
1250	13	3900
1500	15	4400
1750	16.7	5000
2000	18.5	5600

Table 3 gives some data for a series of air-cooled transformers.
1 What is the mass in tonnes of a 800kW transformer?
2 What is the mass in t of a 2000kW transformer?
3 Plot a graph of maximum power, on the x axis, against power loss, on the y axis. Would you say that the graph is a straight line or very slightly curved? Draw a best fit for it.
4 On the same piece of graph paper, and using the same x axis scale, plot a graph of maximum power against mass of transformer. Would you say that the graph is a straight line, slightly curved, or very curved? Draw a best fit.
5 Find the efficiency of the second transformer in the series. Take its maximum power **input** to be 1000kW, and the power loss to be 11kW. (a) What is the power output? (b) What is the efficiency?
6 Find the efficiency of the largest transformer in the series. Take its maximum power **input** to be 2000kW, and the power loss to be 18.5kW. (a) What is the power output? (b) What is the efficiency?
7 What do you notice about the efficiencies as the transformers get larger? Where have you found something similar before? (You could look back at 'Efficiencies of Alternators', pages 36 to 37.)

Development section

A Basic electromagnetism

Some of the magnetic fields which can be produced by electric currents were described on pages 5 to 7, 'Some Magnetic Fields'. You should read this section.

The behaviour of magnetic flux lines

Magnetic flux lines are imaginary. They have been invented to try to describe the behaviour of magnetic fields. But they are a very useful idea, and they are of great help in designing electromagnetic machines.

Some useful facts about the lines are:
1. The arrows on the lines show the direction of the field (the direction in which a small compass needle would point if placed there).
2. Each line forms a closed loop. There are no loose ends to field lines.
3. Lines never cross each other.
4. The lines behave like stretched elastic bands; they always try to shorten themselves. (This is a very useful rule when trying to decide how parts of an electromagnetic device might behave.)
5. The more magnetic flux there is in a region, the closer the flux lines are together; and the stronger is the magnetic field in that region.

Soft and hard magnetic materials

Soft iron is only a very weak magnet, but soft iron can be strongly magnetized by induction when it is placed in a coil of wire which is carrying a current. It can also be magnetized by induction if a permanent magnet is placed near it, or in contact with it. But if the current is switched off, or if the permanent magnet is removed, then the soft iron returns to being only a very weak magnet. It is said to be a **soft magnetic material.**

Silicon alloy steel is also a soft magnetic material, and most industrial electromagnets are made from silicon steel. The cores and yokes of transformers, and the rotors of alternators, are also made from this magnetically soft silicon steel. Its composition is approximately 96% iron and 4% silicon.

Hard magnetic materials are those which, once they have been magnetized, hold their magnetism. Certain alloys based on iron and containing varying percentages of cobalt and nickel are hard magnetic materials. They are cobalt and nickel steels. Hard magnetic materials are used for making permanent magnets.

A machine or a device which needs a magnetic field that goes on and off will need to use a soft magnetic material. A machine or a device that needs a constant magnetic field could use a permanent magnet, made of hard magnetic material.

B Magnetic fields and magnetic particle inspection

Today's inspection jobs

Figure 53 shows some parts which you have been asked to inspect for cracks, by using magnetic fields and magnetic particle inspection. Before doing this you should make sure that you know the shapes of the magnetic fields that are produced by a current in a straight conductor and a current in a coil, pages 5 to 7. You should also read the section on Non-Destructive Testing, pages 8 to 14.

(a) In Figure 53(a) the part which you are to test is a steel cylinder. You have put three turns of thick insulated wire around it. On a sheet of notepaper, sketch the arrangement. Sketch in a current direction, and then some magnetic flux lines. Beside each of A, B and C write in either 'Detected', or 'Not detected', or 'Poorly shown up'.

Figure 53 Testing some components by magnetic particle inspection. (Adapted from Operator Guidance Chart Magnaflux Limited.)

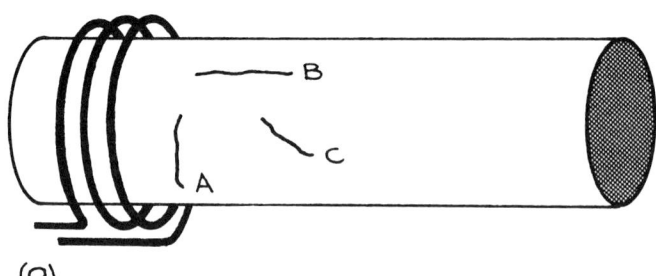

(a)

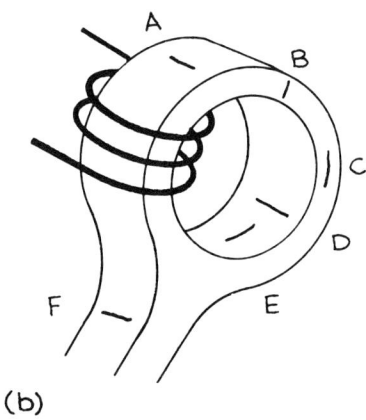

(b)

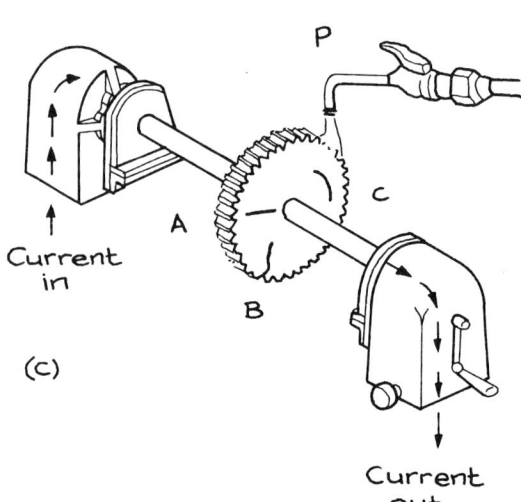

(c)

(b) Figure 53(b) shows the next part which you have been asked to test. Follow the instructions in (a), for the possible cracks A–F.
(c) Figure 53(c) shows a gear which you have been asked to test. The gear is 200mm in diameter. Your instructions tell you to use a current of 2000A to magnetize the gear. What is pipe P for? On a sheet of notepaper, sketch the support for the gear and the gear itself. Mark in the current direction, and then sketch in some magnetic flux lines around the support and in the gear. For the possible cracks, A, B and C label which would be detected well, not at all, or poorly.

C A working solenoid

Some calculations

Solenoids were described in the section on 'Solenoids', pages 15 to 19. You should read this. Let us examine the D3 Model solenoid manufactured by the Warner Electric Company. The solenoid is shown in Figure 2, page 1 and in Figure 22, page 18; have a look at these.

A current I in a conductor of resistance R produces heat at a rate of I^2R. $P = I^2R$. If the current is measured in amps and the resistance in ohms, then the rate of heat production (or power, P) is in watts (J/s). Also,

$$\text{Resistance of a conductor} = \frac{\text{Potential difference between the ends}}{\text{Current in the conductor}}$$

$$R = \frac{V}{I}$$

The solenoid works on a 12V direct current supply. At 20°C, a common atmospheric temperature, the resistance of the coil is 1.2Ω. At 90°C the coil resistance is 1.6Ω.

Question C.1  The solenoid has not been used for several hours, and the atmospheric temperature is 20°C. It is switched on.
(a) What is the current in the coil?
(b) What is the rate of heat production, in W?
(c) What is the rate of heat production in J/s?

C.2 The solenoid is used almost continuously for an hour, and the coil reaches a temperature of 90°C.
(a) What is the current in the coil?
(b) Has the current increased or decreased?
(c) Do you think this will have increased or decreased the strength of the electromagnetic field?
(d) If the plunger is held at some chosen position, will the pull on the plunger increase or decrease as the temperature rises?
(e) Have a look at the graphs in Figure 23. Do these confirm your conclusion?

C.3 What is the rate of heat production in the solenoid at 90°C?

C.4 If the plunger is free to move, then the pull from the magnetic field accelerates the plunger into the coil. Users of solenoids may need to be able to work out forces and accelerations, so that they can calculate how long it will take for the plunger to move a piece of machinery.

One of Newton's laws of motion relates the unbalanced force on a body to the mass and acceleration of the body.

Unbalanced force = Mass × Acceleration
$$F = m \times a$$

The **mass** of the plunger is 0.02kg. The plunger movement is well lubricated. Use Figure 23, page 19, and find answers to the following questions, for the coil at 20°C.

(a) When the plunger is 14mm out (stroke = 14mm) what is the force on the plunger?
(b) What is the acceleration of a free plunger, at 14mm stroke?
(c) At 8mm stroke what is the force on the plunger (to the nearest whole number)?
(d) What is the acceleration of a free plunger at 8mm stroke?
(e) Is the acceleration decreasing, staying the same, or increasing as the plunger moves in?
(f) What is the value of the acceleration of a body falling freely under gravity?
(g) Would you say that, compared with the acceleration due to gravity, the acceleration of the plunger is very small, small, similar, large, or very large?
(h) If the plunger is not held back in any way, do you think that it would hit the base gently or hard?
(i) In practice, what would normally tend to hold the plunger back, and reduce the acceleration?
(j) At a stroke of 14mm, the load which the plunger is moving exerts a backwards pull of 4N on the plunger. What is the unbalanced force pulling the plunger into the coil?
(k) What is the acceleration of the plunger now, at stroke 14mm?

The user and the manufacturer

Clearly, for a particular job just any solenoid will not do. If the magnetic force is too great for the job, the plunger will batter the solenoid to pieces. If the temperature rise becomes too great, the insulation on the wire of the coil will be damaged and the coil will burn out. The user needs to tell the manufacturer what the solenoid has to do. For instance, it might need to have a stroke of 40mm, develop a pull of 50N when fully closed, operate on a 24V a.c. supply, work on a machine where the temperature is between 30°C and 40°C, and be on for 15 seconds in every full minute. The manufacturer's engineers will then either recommend one of their existing solenoids which closely meets these needs, or they may offer to design and make a model specially.

If the manufacturer is going to make a new model, the engineers will work out the amount of iron in the plunger, and the magnetic flux that will be needed. Then they can work out the number of turns in the coil, and the current. From that they can work out the resistance of the coil, and so the cross-section area that will be needed for the wire. They can work out how much heat will be produced in the 15 seconds that the solenoid will operate. They can calculate whether this heat can escape by conduction, convection and radiation in the 45 seconds that the device will be off, so that the temperature will not rise to an unsafe level.

Then the separate parts can be made and assembled. The new solenoid can be checked, and tested, using electrical instruments, thermometers, and a newton meter.

Figure 3 and Figure 4 show a large concrete-making plant. Several of the moving parts are operated by pistons moved by compressed air; and the valves which control the compressed air are operated by solenoids.

D Alternating current generators

Basic electromagnetic induction

A way in which electromagnetic induction can produce a current in a circuit was described on pages 32 to 39. You should read these pages.

Look at Figure 39, page 33, in particular at the fixed coil and at the terminals A and B at the ends of the coil.

1. If the magnetic flux through the coil changes, then an e.m.f. is generated in the coil.
2. The flux throught he coil can be changed by rotating the magnetic field.
3. The size of the e.m.f. is proportional to:
   (a) The strength of the magnetic field.
   (b) The rate at which the field rotates.
   (c) The number of turns of wire in the coil.
4. If the coil is connected to a circuit, then the e.m.f. will produce a current in the circuit.
5. Rotating the magnetic field through one complete revolution produces one complete cycle of alternating current in the circuit (Figure 39 and Figure 40).
6. If one complete cycle takes 1 second, then the frequency is 1Hz. If there are 20 complete cycles per second, the frequency is 20Hz.

Some calculations

Question D.1

Table 4 is for a simple fixed-coil alternator like that in Figure 39, page 33. It shows the e.m.f. generated at the coil terminals for three variations: in the strength of the field, the number of turns in the coil and the rate of rotation. From your knowledge of how these affect the e.m.f., give the e.m.f.s for arrangements (a)–(i).

Table 4 Some information about a fixed-coil, rotating-field alternator. Table 4 is for a simple fixed-coil alternator like that in Figure 39, page 33.

	Magnetic field strength (field strength units)	Number of turns on coil	Rate of rotation /Hz	e.m.f generated at coil terminals/V
	1	1	10	2
	2	1	10	4
	2	1	20	8
(a)	3	1	10	–
(b)	4	1	10	–
(c)	1	2	10	–
(d)	1	3	10	–
(e)	1	1	20	–
(f)	1	1	30	–
(g)	1	1	50	–
(h)	2	2	10	–
(i)	2	2	20	–

D.2 Frequency and engine speed

In Figure 38, the engine shaft is connected to a magnet (in fact an electromagnet) which produces a field like that in Figure 39, page 33. The engine can be adjusted to run at the right speed to produce alternating current of frequency 50Hz for use in the United Kingdom, and of frequency 60Hz for use in the USA.

(a) In the UK, what must the engine speed be in **1)** rev/s, **2)** rev/min?
(b) In the USA, what must the engine speed be in **1)** rev/s **2)** rev/min?
(c) The alternator is designed and made with a voltage regulator. Why would this be needed in (b)?

D.3 Efficiencies of alternators

The efficiencies of alternators was described on pages 36 to 37. You should read these pages.

Table 5 gives information about a series of three generators, with the same basic design. What are the efficiencies of the three alternators? what trend do you notice?

Table 5 Power information for some small mobile electricity generators.

Generator model	Diesel engine output power/kW	Alternator output power/kW	Efficiency/%
A	8	6	
B	10	8	
C	12	10	

D.4 Power, current and output voltage

The power which is being supplied to an electric circuit is:

Power = Current × Potential difference
$P = I \times V$
$P = IV$

Have a look at the generator in Figure 38, page 32. It can deliver a maximum of 5kW of power, at either 220V or 110V.
(a) The alternator is delivering 4.4kW (4400W) of power to do work.
  1 If the output voltage is 220V, what is the current?
  2 If the output voltage is 110V, what is the current?
(b) The alternator is delivering its full power of 5kW to do work.
  1 If the output voltage is 220V, what is the current?
  2 If the output voltage is 110V, what is the current?

Getting direct current for the rotor coil

In the 5kW alternator, the rotating magnetic field was produced by a rotating electromagnet. This electromagnet needs direct current. Have a look back at Figure 41, page 34, which shows the coils and the poles of the rotor. How can direct current be supplied for the rotor? It could be obtained from a battery, like a car battery. But a suitable battery would be very heavy. Also, it would need recharging. It would require maintenance; and it would probably wear out and need replacing long before any other major part of the alternator. A battery is not a suitable or convenient way of providing the direct current.

Another source of electric power would be the alternator itself. But the alternator produces alternating current, so a current rectifier is needed, to turn some of the alternating current into direct current. A **diode** is used.

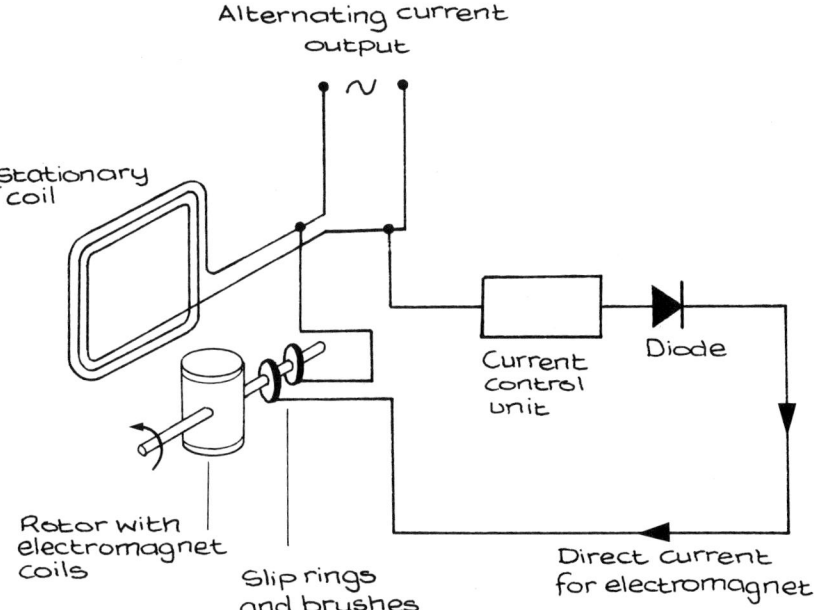

Figure 54 This is a simple circuit for an alternator. The electromagnet in the rotor needs direct current. This diagram shows how the direct current can be obtained from the alternating current output.

Figure 54 shows the circuit. On the left is the stationary coil. Connected in parallel with this is the rotor circuit. Look at the righthand output wire of the stationary coil; a wire goes to a **current control unit** and then to a **diode**. The diode allows current through only in the forwards direction; and so **direct current** goes to the rotor. From there it goes to the other output wire of the stationary coil, so completing the circuit (through the stationary coil).

The current control unit controls the size of the current through the rotor coil; and so it controls the amount of magnetic flux which the electromagnet produces. The rotor coil uses about 200W to 300W of power to produce the magnetic flux. This is 0.2kW to 0.3kW, compared with the 5kW output (4% to 6% of the total power). The arrangement is neat, very convenient and very reliable.

Very large alternators work on the same principle. (A possible problem may occur to you. If the rotor is stationary, is there any current through the fixed coils? If not, is there any magnetic flux due to **electro**magnetism in the rotor? If not, then how can an e.m.f. be induced in the stationary coil when the rotor is turned around? The answer is that the steel core of the rotor remains as a weak magnet. When the rotor is first turned around, the small amount of magnetic flux is enough to generate a small e.m.f. in the stationary coil. Then a small current goes through the rotor coil. What happens then? What then very rapidly happens?

E Transformers

Why transformers are needed, the way in which they work, and their efficiency were described on pages 39 to 46. Look back at this, and read the pages again.

A summary of some of the main points on those pages is given here.

Transformers are used to change the voltage of an electrical power supply.

They do this by electromagnetic induction.

A current in the primary coil produces a magnetic flux, and this flux is shared with the secondary coil.

A changing current in the primary coil (alternating current) produces a changing flux.

The changing flux induces a current in the secondary coil if it is in a complete circuit.

The current in the secondary coil and its complete circuit is alternating current.

Power is transferred from the primary circuit to the secondary circuit, even though these are not in electrical contact with each other. (The wires are insulated.)

The transfer of power is brought about by the changing magnetic flux.

$$\frac{\text{Number of primary turns}}{\text{Number of secondary turns}} = \frac{\text{Primary volts}}{\text{Secondary volts}}$$

$$\frac{N_P}{N_S} = \frac{V_P}{V_S}$$

Input volts, output volts and coils turns

Question E.1

(a) A step-down transformer is to use a 440V supply to the primary, to give a 110V output from the secondary. The primary coil is to have 160 turns in it. How many turns must be put in the secondary coil?

(b) A stepdown transformer is to use a 240V supply to the primary, to provide a 20V output from the secondary. The primary coil is to have 72 turns on it. How many turns must the secondary coil have?

(c) A transformer is needed to provide power for 110V tools from the 415V mains. The primary coil is to have 385 turns. How many turns must be put on the secondary coil?

Input current and output current

Figure 55 shows a transformer which is taken as being 100% efficient. Four kilowatts of power is being supplied to the primary coil; and 4kW of power is being taken from the secondary coil. The transformer is a 400V/100V step-down transformer. Now,

$$\begin{aligned}\text{Power} &= \text{Potential difference} \times \text{Current}\\P &= V \times I\\\text{watts} &= \text{volts} \times \text{amps}\end{aligned}$$

From this equation we can calculate the current in the primary coil, and the current in the secondary coil.

For the **primary coil**

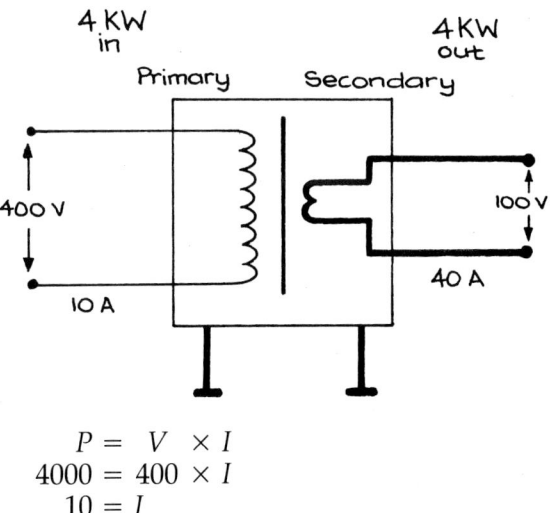

Figure 55 This shows a 100% efficient transformer. The output power equals the input power.

$$\begin{aligned}P &= V \times I\\4000 &= 400 \times I\\10 &= I\end{aligned}$$

and the primary current is 10A.

For the **secondary coil**

$$\begin{aligned}P &= V \times I\\4000 &= 100 \times I\\40 &= I\end{aligned}$$

and the secondary current is 40A.
The transformer is a 4:1 step-down transformer.

The **voltage** has gone **down** to ¼ of its original value.
The **current** has gone **up** to four times its original value.

In a transformer that steps down the voltage, the secondary current is always greater than the primary current, so the cross-section area of the secondary conductor has to be greater than that of the primary conductor. Figure 55 shows this, and so does Figure 56.

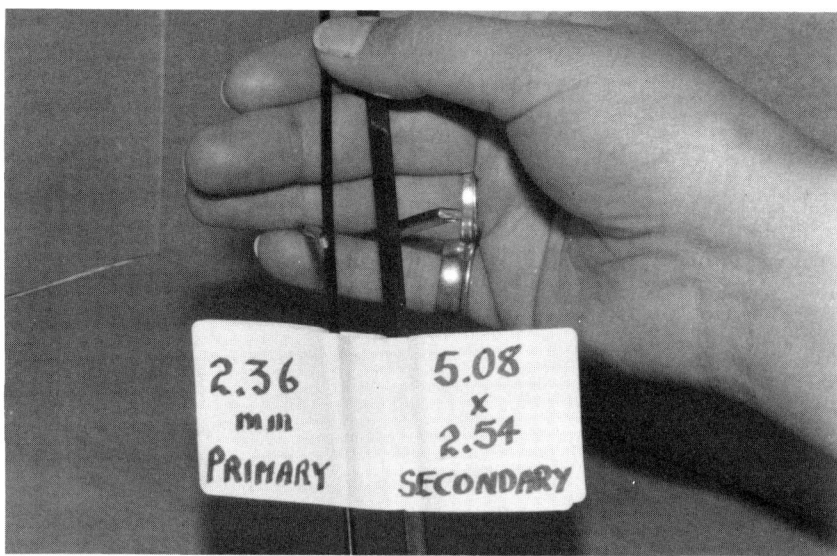

Figure 56 This photograph shows the primary and secondary conductors for a 3 kW, 240V: 110V step-down transformer. (Loheat Limited, UK.)

E.2 100% efficient transformers

The following transformers are taken as being 100% efficient, and they are providing the power which is shown. Calculate the current in **1)** the primary winding, and **2)** the secondary winding for each.
(a) 6kW; 200V/100V.
(b) 5kW; 250V/50V.
(c) 10kW; 415V/110V.

We have seen that in a step-down transformer the current in the secondary coil will be greater than the current in the primary coil. In the secondary if the **voltage goes down,** the **current goes up,** compared with the primary. The cross-section areas of the conductors must be right for this.

Let us look at two examples. Have a look back at Figure 52. The technician is completing a 2kW step-down transformer. Some of the transformer details are shown in Table 6.

Table 6 Some transformer details. (Loheat Limited, UK).

Coil	Voltage/V	Turns	Conductor cross-section	Resistance/Ω
Primary	240	120	2.6 mm diameter wire	0.18
Secondary	110	55	Two 2.8 mm diameter wires joined in parallel	0.043

In the secondary winding, two 2.8mm diameter wires are used instead of one thicker wire. This is because the thicker wire would be too stiff to bend round the coil frame without damaging the wire and its insulation.

Figure 56 shows the conductors for a 3kW, 240V/110V transformer. The cross-section of the primary conductor is circular, 2.36mm diameter, and that of the secondary conductor is rectangular, 5.08mm × 2.54mm. A thin broad copper strip can be wound easily, but a thick cylindrical copper wire cannot.

Laminations

Have a look again at Figure 52. The technician is finishing making the core and yoke. They are made from many thin sheets of magnetically soft steel. This one is in the shape of an E. At the far side a straight strip in the shape of an I will be put in place, to close the E. These thin sheets are called **laminations**. Each is coated with a thin layer of insulator, usually varnish.

The changing flux induces a current in a secondary coil circuit. It also induces currents in the steel core and yoke. These currents heat the steel, and waste energy. It has been found that if the core and yoke are made of thin sheets of steel, each insulated from the others, then the currents are much smaller than for a solid core and yoke, the heating is much smaller, and the energy loss is much smaller.

The currents are called **eddy currents**. They are formed in other types of alternating current machinery, too. Do you think they might occur in the 5kW a.c. generator, in pages 32 to 38? The stationary coils have to be wound onto something to hold them in place. The coils are wound onto a magnetically soft steel frame. The frame is made up of laminations. Does this give a clue?

Answers

C1 (a) 10A (b) 120W (c) 120J/s
C2 (a) 7.5A
C3 90W
C4 (a) 6N (b) 300m/s^2 (c) 10N (d) 500m/s^2 (f) 10m/s^2 (j) 2N (k) 100m/s^2

D1 (a) 6V (b) 8V (c) 4V (d) 6V (e) 4V (f) 6V (g) 10V (h) 8V (i) 16V
D2 (a) **1)** 50rev/s **2)** 3000rev/min (b) **1)** 60rev/s **2)** 3600rev/min
D3 A, 75%; B, 80%; C, 83.3%
D4 (a) **1)** 20A **2)** 40A (b) **1)** 22.7A **2)** 45.5A

E1 (a) 40 turns (b) 6 turns (c) 102 turns
E2 (a) **1)** 30A **2)** 60A (b) **1)** 20A **2)** 100A (c) **1)** 24.1A **2)** 90.9A

INDEX

alternating current 32–9, 51–4
   generators 51
alternator 32–9
armature 21, 24, 26

bimetal strip 16–17
brakes, electromagnetic 21–2, 25–31
brushes 27, 35

circuit breaker 15–17
clutches, electromagnetic 22–5
cooling
   alternators 35–6, 37–8
   brakes 28
   clutches 28
   transformers 44

defects, in materials 9
demagnetisation 14–15
diode 53–4

eddy currents 57
efficiency
   of alternators 36–7, 52
   of transformers 44
electromagnet 7, 8, 12, 24–8, 34
engineers and electromagnetism 1

fields, magnetic 5–8, 10–11, 13–14, 33, 40–1, 48–9
flux lines, magnetic 6–7, 10–11, 12–13, 33, 40–1, 47

heating effect of a current 35–7

induction, electromagnetic 33, 40–1, 51–2

laminations 57

magnet, bar 6
   horseshoe 7
magnetic fields 5–8
magnetic materials 47–8
magnetic particle inspection 8–15, 48–9

NDT, non-destructive testing 8–15, 48–9
noise, in transformers 44–6

relays 1, 19–22
rotor 34, 53–54

slip ring 27, 34–5
solenoids 1, 6, 15–19, 49–51

transformers 39–47, 54–7